Contents

Use of guidance

THE APPROVED DOCUMENTS

This document is one of a series that has been approved and issued by the Secretary of State for the purpose of providing practical guidance with respect to the requirements of Schedule 1 to and regulation 7 of the Building Regulations 2000 (SI 2000/2531) for England and Wales. SI 2000/2531 has been amended by the Building (Amendment) Regulations 2001 (SI 2001/3335)

At the back of this document is a list of all the documents that have been approved and issued by the Secretary of State for this purpose.

Approved Documents are intended to provide guidance for some of the more common building situations. However, there may well be alternative ways of achieving compliance with the requirements. **Thus there is no obligation to adopt any particular solution contained in an Approved Document if you prefer to meet the relevant requirement in some other way.**

Other requirements

The guidance contained in an Approved Document relates only to the particular requirements of the Regulations which the document addresses. The building work will also have to comply with the requirements of any other relevant paragraphs in Schedule 1 to the Regulations.

There are Approved Documents which give guidance on each of the Parts of Schedule 1 and on regulation 7.

LIMITATION ON REQUIREMENTS

In accordance with regulation 8, the requirements in Parts A to K and N (except for paragraphs H2 and J6) of Schedule 1 to the Building Regulations do not require anything to be done except for the purpose of securing reasonable standards of health and safety for persons in or about buildings (and any others who may be affected by buildings or matters connected with buildings).

Paragraphs H2 and J6 are excluded from Regulation 8 because they deal directly with prevention of the contamination of water. Parts L and M are excluded because they respectively address the conservation of fuel and power and access facilities for disabled people. These matters are amongst the purposes, other than health and safety, that may be addressed by Building Regulations.

MATERIALS AND WORKMANSHIP

Any building work which is subject to the requirements imposed by Schedule 1 to the Building Regulations should, in accordance with regulation 7, be carried out with proper materials and in a workmanlike manner.

You may show that you have complied with regulation 7 in a number of ways. These include the appropriate use of a product bearing CE marking in accordance with the Construction Products Directive (89/106/EEC)[1] as amended by the CE Marking Directive (93/68/EEC)[2] , or a product complying with an appropriate technical specification (as defined in those Directives), a British Standard, or an alternative national technical specification of any state which is a contracting party to the European Economic Area which, in use, is equivalent, or a product covered by a national or European certificate issued by a European Technical Approval issuing body, and the conditions of use are in accordance with the terms of the certificate. You will find further guidance in the Approved Document supporting regulation 7 on materials and workmanship.

Independent certification schemes

There are many UK product certification schemes. Such schemes certify compliance with the requirements of a recognised document which is appropriate to the purpose for which the material is to be used. Materials which are not so certified may still conform to a relevant standard.

Many certification bodies which approve such schemes are accredited by UKAS.

Technical specifications

Under section 1(a) of the Building Act, Building Regulations may be made for various purposes including health, safety, welfare, convenience, conservation of fuel and power and prevention of contamination of water. Standards and technical approvals are relevant guidance to the extent that they relate to these considerations. However, they may also address other aspects of performance such as serviceability, or aspects, which although they relate to the purposes listed above, are not covered by the current Regulations.

1 As implemented by the Construction Products Regulations 1991 (SI 1991/1620).

2 As implemented by the Construction Products (Amendment) Regulations 1994 (SI 1994/3051).

When an Approved Document makes reference to a named standard, the relevant version of the standard is the one listed at the end of the publication. However, if this version has been revised or updated by the issuing standards body, the new version may be used as a source of guidance provided it continues to address the relevant requirements of the Regulations.

The appropriate use of a product which complies with a European Technical Approval as defined in the Construction Products Directive will meet the relevant requirements.

The Department intends to issue periodic amendments to its Approved Documents to reflect emerging harmonised European Standards. Where a national standard is to be replaced by a European harmonised standard, there will be a co-existence period during which either standard may be referred to. At the end of the co-existence period the national standard will be withdrawn.

THE WORKPLACE (HEALTH, SAFETY AND WELFARE) REGULATIONS 1992

The Workplace (Health, Safety and Welfare) Regulations 1992 contain some requirements which affect building design. The main requirements are now covered by the Building Regulations, but for further information see: *Workplace health, safety and welfare. Workplace (Health, Safety and Welfare) Regulations 1992. Approved Code of Practice L24.* Published by HSE Books 1992 (ISBN 0 7176 0413 6).

The Workplace (Health, Safety and Welfare) Regulations 1992 apply to the common parts of flats and similar buildings if people such as cleaners and caretakers are employed to work in these common parts. Where the requirements of the Building Regulations that are covered by this Part do not apply to dwellings, the provisions may still be required in the situations described above in order to satisfy the Workplace Regulations.

MIXED USE DEVELOPMENT

In mixed use developments part of a building may be used as a dwelling while another part has a non-domestic use. In such cases, if the requirements of this Part of the Regulations for dwellings and non-domestic use differ, the requirements for non-domestic use should apply in any shared parts of the building.

Summary guide to the use of this Approved Document

Routes to compliance for non-domestic buildings

STEP	TEST		ACTION
DESIGN			
	Choose method of compliance		
	Elemental		Go to 1
	Whole building method		Go to 25
	Carbon emissions calculation method		Go to 30
Compliance by Elemental method			
1	Are all U-values ≤ the corresponding values from Table 1?	No Yes	Revise U-values and repeat 1 or Go to 3 Continue
2	Are the areas of openings ≤ the corresponding values in Table 2?	Yes No	Go to 4 Reduce opening areas and repeat 2 or continue
3	Is the average U-value ≤ to that of a notional building of the same size and shape as described in paragraphs 1.15-1.16 and taking into account the heating system efficiency as described in para 1.32?	No	FAIL - revise design and go to 1 or go to START and test compliance by another route
4	Do all occupied spaces satisfy the solar overheating criteria in para 1.20 et seq ?	No Yes	Adjust window areas or shading provisions Continue
5	Does any centralised heating plant as described in paragraph 1.25 meet the carbon intensity criteria of Table 5?	Yes No	Go to 7 Select different heating plant and repeat 5 or continue
6	Is the average U-value $\leq U_{ref} \cdot \frac{\varepsilon_{ref}}{\varepsilon_{act}}$ from paragraph 1.32?	No Yes	**FAIL** – revise design and go to 1 or go to **START** and test compliance by another route Continue
7	Do the heating system controls comply with paragraphs 1.33 and 1.34?	No Yes	**FAIL** – revise controls and repeat 7 or go to **START** and test compliance by another route Continue
8	Does the HWS system and the associated controls comply with paragraphs 1.35 to 1.37?	No Yes	**FAIL** – revise HWS system and controls and repeat 8 or go to **START** and test compliance by another route Continue
9	Does the insulation to pipes, ducts and vessels comply with paragraphs 1.38 to 1.40?	No Yes	**FAIL** – revise insulation specification and repeat 9 or go to **START** and test compliance by another route Continue
10	Is there general or display lighting serving more than 100m^2?	No Yes	Go to 19 Continue
11	Is the building an office, industrial or storage building?	No Yes	Go to 14 Continue
12	Is the average luminaire-lumens/circuit watt ≥ 40?	No Yes	Revise lighting design and repeat 12 or go to **START** and test compliance by another route Continue
13	Do the lighting controls comply with paragraphs 1.56 and 1.57?	Yes No	Go to 16 Revise the controls and repeat 13 or go to **START** and test compliance by another route
14	Is the average lamp plus ballast efficacy ≥ 50 lamp-lumens per circuit watt?	No Yes	Revise lighting design and repeat 14 or go to **START** and test compliance by another route Continue
15	Do the lighting controls meet the guidance in paragraph 1.58	No Yes	Revise the controls and repeat 15 or go to **START** and test compliance by another route Continue

Routes to compliance for non-domestic buildings Cont.

STEP	TEST		ACTION
16	Is there any display lighting?	No Yes	Go to 19 Continue
17	Does any display lighting have an average lamp plus ballast efficacy ≥ 15 lamp-lumens per circuit watt?	No Yes	Revise display lighting and repeat 17 or go to **START** and test compliance by another route Continue
18	Do the display lighting controls meet the standards of paragraph 1.59?	No Yes	Revise display lighting controls and repeat 18 or go to **START** and test compliance by another route Continue
19	Does the building have any air conditioning or mechanical ventilation systems that serve more than 200 m² floor area?	No Yes	Go to 23 Continue
20	Is it an office building?	No Yes	Go to 22 Continue
21	Is the Carbon Performance Rating ≤ the values in Table 11	No Yes	Revise the ACMV design and repeat 21 or go to **START** and test compliance by another route Go to 23
22	Is the specific fan power ≤ the values given in paragraph 1.67?	No Yes	Revise design of mechanical ventilation system and repeat 22 or go to **START** and test compliance by another route Continue
23	Are there any sun-spaces?	No Yes	**PASS Elemental Method** and go to 34 Continue
24	Do the sun-spaces meet the criteria of paragraphs 1.77 to 1.79?	No Yes	Revise sun-space design and repeat 24 or go to **START** and test compliance by another route **PASS Elemental Method** – go to 34
Compliance by Whole Building method			
25	Select building type	Office: School: Hospital: Other:	Go to 26 Go to 28 Go to 29 Method not suited - go to **START** and test compliance by another route
26	Is the whole office CPR ≤ the relevant value in Table 12	No Yes	**FAIL** – Revise design and repeat or go to **START** and test compliance by another route Continue
27	Are the proposed building fabric performances no worse than those given in table 3 and paragraphs 1.9-1.11 and 1.17-1.19 respectively	No Yes	Revise details and repeat 28 or go to **START** and test compliance by another route **PASS Whole Building Method** - Go to 34
28	Does the school meet the requirements of DfEE Building bulletin 87?	No Yes	**FAIL** - Revise design and repeat 28 or go to **START** and test compliance by another route **PASS Whole Building Method** - Go to 34
29	Does the hospital meet the requirements of NHS Estates guidance?	No Yes	**FAIL** - Revise design and repeat 29 or go to **START** and test compliance by another route **PASS Whole Building Method** – Go to 34
Compliance by Carbon Emission Calculation Method			
30	Does the notional building meet the standards in paragraph 1.75?	No Yes	Revise notional design and repeat 30 or go to **START** and test compliance by another route Continue
31	Does the envelope of the proposed building meet the standards of paragraph 1.75(b)	No Yes	Revise proposed building envelope and repeat 31 or go to **START** and test compliance by another route Continue
32	Has the calculation method been agreed as appropriate to the application (paragraph 1.76)?	No Yes	**FAIL** - go to **START** and test compliance by another route Continue
33	Is the carbon emitted by the proposed building ≤ that emitted by the notional building?	No Yes	**FAIL** – revise design and repeat 33 or go to **START** and test compliance by another route **PASS Carbon Emissions Calculation Method** go to 34

Routes to compliance for non-domestic buildings Cont.

STEP	TEST		ACTION
CONSTRUCTION			
34	Is the building control body reasonably convinced that the fabric insulation in the actual building is continuous ? (paragraph 2.1)	No	**FAIL** – carry out remedial work and repeat 34
		Yes	Continue
35	Is the building control body reasonably convinced that the building is satisfactorily airtight ? (paragraph 2.2)	No	Identify leaks, seal and re-test to meet standards of paragraph 2.4.
		Yes	Continue
36	Has inspection and commissioning been completed satisfactorily? (paragraphs 2.5 and 2.6)	No	Complete commissioning and repeat 36
		Yes	**PASS Construction is satisfactory** – continue
PROVIDING INFORMATION			
37	Has the log-book been prepared (paragraphs 3.1 and 3.2)?	No	Prepare log-book and repeat 37
		Yes	Continue
38	Has a metering strategy been prepared and sufficient meters and sub-meters installed? (paragraphs 3.3 et seq)	No	Produce strategy / install meters and sub-meters and repeat 38
		Yes	**BUILDING COMPLIES**

The Requirements

CONSERVATION OF FUEL AND POWER THE REQUIREMENT L2

This Approved Document, which takes effect on 1 April 2002, deals with the following Requirements which are contained in the Building Regulations 2000 (as amended by SI 2001/3335).

Requirement	*Limits on application*
Buildings or parts of buildings other than dwellings	
L2. Reasonable provision shall be made for the conservation of fuel and power in buildings or parts of buildings other than dwellings by –	
(a) limiting the heat losses and gains through the fabric of the building;	
(b) limiting the heat loss:	
(i) from hot water pipes and hot air ducts used for space heating;	
(ii) from hot water vessels and hot water service pipes;	
(c) providing space heating and hot water systems which are energy-efficient;	
(d) limiting exposure to solar overheating;	
(e) making provision where air conditioning and mechanical ventilation systems are installed, so that no more energy needs to be used than is reasonable in the circumstances;	Requirements L2(e) and (f) apply only within buildings and parts of buildings where more than 200m^2 of floor area is to be served by air conditioning or mechanical ventilation systems.
(f) limiting the heat gains by chilled water and refrigerant vessels and pipes and air ducts that serve air conditioning systems;	
(g) providing lighting systems which are energy-efficient;	Requirement L2(g) applies only within buildings and parts of buildings where more than 100 m^2 of floor area is to be served by artificial lighting.
(h) providing sufficient information with the relevant services so that the building can be operated and maintained in such a manner as to use no more energy than is reasonable in the circumstances.	

OTHER CHANGES TO THE BUILDING REGULATIONS 2000

Attention is particularly drawn to the following changes to the provisions of the Building Regulations 2000 made by the Building (Amendment) Regulations 2001.

Amendment to Regulation 2 – Interpretation

The definition of controlled service or fitting in Regulation 2(1) is amended to:-

"controlled service or fitting means a service or fitting in relation to which Part G, H, J or L of Schedule 1 imposes a requirement;"

Amendments to Regulation 3 – Meaning of building work

Regulation 3(1) is amended as follows:

"3(1) ...

(b) subject to paragraph (1A), the provision or extension of a controlled service or fitting in or in connection with a building;"

New Regulation 18 – Testing of building work

A new Regulation 18 is subtituted which says:-

"**Testing of building work**

The local authority may make such tests of any building as may be necessary to establish whether it complies with regulation 7 or any of the applicable requirements contained in Schedule 1."

Section 0: General guidance

Performance

0.1 In the Secretary of State's view requirement L2 (a) will be met by the provision of energy efficiency measures which:

a) limit the heat loss through the roof, wall, floor, windows and doors etc by suitable means of insulation, and where appropriate permit the benefits of solar heat gains and more efficient heating systems to be taken into account; and

b) limit the heat gains in summer; and

c) limit heat losses (and gains where relevant) through unnecessary air infiltration by providing building fabric which is reasonably airtight.

0.2 In the Secretary of State's view requirement L2 (b) will be met by limiting the heat loss from hot water pipes and hot air ducts used for space heating and from hot water vessels and hot water service pipes by applying suitable thicknesses of insulation where such heat does not make an efficient contribution to the space heating.

0.3 In the Secretary of State's view requirement L2 (c) will be met by the provision of space heating and hot water systems with reasonably efficient equipment such as heating appliances and hot water vessels where relevant, and timing and temperature controls, and suitable energy consumption metering, that have been appropriately commissioned such that the heating and hot water systems can be operated effectively as regards the conservation of fuel and power.

0.4 In the Secretary of State's view requirement L2 (d) will be met by the appropriate combination of passive measures, such as limiting the area of glazing which is not shaded and providing external building fabric that limits and delays heat penetration, with active measures, such as night ventilation, so that the effects of solar heat gains are kept within limits that are reasonable in the circumstances.

0.5 In the Secretary of State's view, when buildings are proposed to be mechanically ventilated or air conditioned, requirement L2 (e) will be met by:-

a) limiting the demands from within the building for heating and cooling and circulation of air, water and refrigerants; and

b) providing reasonably efficient plant and distribution systems, and timing, temperature and flow controls, and suitable energy consumption metering, that have been appropriately commissioned.

0.6 In the Secretary of State's view requirement L2 (f) will be met by limiting the heat gains to chilled water and refrigerant vessels and pipes and air ducts by applying suitable thicknesses of insulation including vapour barriers.

0.7 In the Secretary of State's view requirement L2 (g) will be met by the provision of lighting systems that where appropriate:

a) utilise energy-efficient lamps and luminaires, and

b) have suitable manual switching or automatic switching, or both manual and automatic switching controls, and

c) have suitable energy consumption metering, and

d) have been appropriately commissioned.

0.8 In the Secretary of State's view requirement L2 (h) will be met by providing information, in a suitably concise and understandable form, and including the results of performance tests carried out during the works, that shows how the building and its relevant building services can be operated and maintained so that they use no more energy than is reasonable in the circumstances.

Introduction to the Provisions

Technical Risk

0.9 Guidance on avoiding technical risks (such as rain penetration, condensation etc) which might arise from the application of energy conservation measures is given in BRE Report No 262: "Thermal Insulation: avoiding risks", 2002 Edition. As well as giving guidance on ventilation for health, Approved Document F contains guidance on the provision of ventilation to reduce the risk of condensation in roof spaces. Approved Document J gives guidance on the safe accommodation of combustion systems including the ventilation requirements for combustion and the proper working of flues. Approved Document E gives guidance on achieving satisfactory resistance to the passage of sound. Guidance on some satisfactory design details is given in the robust details publication [1].

1 Limiting thermal bridging and air leakage: Robust construction details for dwellings and similar buildings; TSO, 2001

Thermal conductivity and transmittance

0.10 Thermal conductivity (i.e. the lambda-value) of a material is a measure of the rate at which that material will pass heat and is expressed in units of Watts per metre per degree of temperature difference (W/mK).

0.11 Thermal transmittance (i.e. the U-value) is a measure of how much heat will pass through one square metre of a structure when the air temperatures on either side differ by one degree. U-values are expressed in units of Watts per square metre per degree of temperature difference (W/m^{2}K).

0.12 Exposed element means an element exposed to the outside air (including a suspended floor over a ventilated or unventilated void, and elements so exposed indirectly via an unheated space), or an element in the floor or basement in contact with the ground. In the case of an element exposed to the outside air via an unheated space (previously known as a "semi-exposed element") the U-value should be derived from the transmission heat loss coefficient[2]. Party walls, separating two premises that can reasonably be assumed to be heated to the same temperature, are assumed not to need thermal insulation.

0.13 In the absence of test information, thermal conductivities and thermal transmittances (U-values) may be taken from the tables in this Approved Document or alternatively in the case of U-values they may be calculated. However, if test results for particular materials and makes of products obtained in accordance with a harmonised European Standard are available they should be used in preference. Measurements of thermal conductivity should be made according to BS EN 12664[3], BS EN 12667[4], or BS EN 12939[5]. Measurements of thermal transmittance should be made according to BS EN ISO 8990[6] or, in the case of windows and doors, BS EN ISO 12567-1[7]. The size and configuration of windows for testing or calculation should be representative of those to be installed in the building, or conform to published guidelines on the conventions for calculating U-values[8].

U-value reference tables

0.14 Appendix A contains tables of U-values and examples of their use, which provide a simple way to establish the amount of insulation needed to achieve a given U-value for some typical forms of construction. They yield cautious results that, in practice, are equal or better than the stated U values. However specific calculations where proprietary insulation products are proposed may indicate that somewhat less insulation could be reasonable. The values in the tables have been derived taking account of typical thermal bridging where appropriate. Appendix A also contains tables of indicative U-values for windows, doors and rooflights.

Calculation of U-values

0.15 U-values should calculated using the methods given in:

- for walls and roofs: BS EN ISO 6946[9]
- for ground floors: BS EN ISO 13370[10]
- for windows and doors: BS EN ISO 10077-1[11] or prEN ISO 10077-2[12]
- for basements: BS EN ISO 13370 or the BCA/NHBC Approved Document.[13]

---

2 BS EN ISO 13789:1999 Thermal performance of buildings - Transmission heat loss coefficient - Calculation method

3 BS EN 12664:2001 Thermal performance of building materials and products – Determination of thermal resistance by means of guarded hot plate and heat flow meter methods – Dry and moist products of low and medium thermal resistance

4 BS EN 12667:2000 Thermal performance of building materials and products – Determination of thermal resistance by means of guarded hot plate and heat flow meter methods – Products of high and medium thermal resistance

5 BS EN 12939:2001 Thermal performance of building materials and products – Determination of thermal resistance by means of guarded hot plate and heat flow meter methods – Thick products of high and medium thermal resistance

6 BS EN ISO 8990:1996 Thermal insulation – Determination of steady-state thermal transmission properties – Calibrated hot box

7 BS EN ISO 12567-1:2000 Thermal performance of windows and doors – Determination of thermal transmittance by hot box method – Part 1: Complete windows and doors

8 Conventions for the calculation of U-values, BRE: expected publication date early 2002

9 BS EN ISO 6946:1997 Building components and building elements – Thermal resistance and thermal transmittance – Calculation method

10 BS EN ISO 13370:1998 Thermal performance of buildings – Heat transfer via the ground – Calculation methods

11 BS EN ISO 10077-1:2000 Thermal performance of windows, doors and shutters – Calculation of thermal transmittance – Part 1: Simplified methods

12 prEN ISO 10077-2: Thermal performance of windows, doors and shutters – Calculation of thermal transmittance – Part 2: Numerical method for frames

13 Approved Document "Basements for dwellings", ISBN 0-7210-1508-5, 1997

For building elements not covered by these documents the following may be appropriate alternatives: BRE IP 5/98[14] for curtain walling the CAB Guide[15] or the methodology in CWCT guidance[16,17], BRE guidance for light steel frame walls[18], or finite element analysis in accordance with BS EN ISO 10211-1[17] or BS EN ISO 10211-2[20] . BRE guidance on conventions for establishing U-values[8] can be followed. Some examples of U-value calculations are given in Appendix B and Appendix C gives data for ground floors and basements.

0.16 Thermal conductivity values for common building materials can be obtained from BS EN 12524[21] or the CIBSE Guide Section A3[22], but for ease of reference a table of common materials is given in Appendix A. For specific insulation products, data should be obtained from the manufacturers.

0.17 When calculating U-values, the thermal bridging effects of, for instance, timber joists, structural and other framing, normal mortar bedding and window frames should generally be taken into account using the procedure given in BS EN ISO 6946 (some examples are given in Appendix B). Thermal bridging can be disregarded however where the difference in thermal resistance between the bridging material and the bridged material is less than 0.1 m^2K/W. For example normal mortar joints need not be taken into account in calculations for brickwork. Where, for example, walls contain in-built meter cupboards, and ceilings contain loft hatches , recessed light fittings, etc, area-weighted average U-values should be calculated.

Roof window

0.18 A roof window is a window in the plane of a pitched roof and may be considered as a rooflight for the purposes of this Approved Document.

Basis for calculating areas

0.19 The dimensions for the areas of walls, roofs and floors should be measured between finished internal faces of the external elements of the building including any projecting bays. In the case of roofs they should be measured in the plane of the insulation. Floor areas should include non-useable space such as builders' ducts and stairwells.

Air permeability

0.20 Air permeability is the physical property used to quantify airtightness of building fabric. It measures the resistance of the building envelope to inward or outward air permeation. It is defined as the average volume of air (in cubic metres per hour) that passes through unit area of the structure of the building envelope (in square metres) when subject to an internal to external pressure difference of 50 Pa. It is expressed in units of cubic metres per hour, per square metre of envelope area, at a pressure difference of 50 Pa. The envelope area of the building is defined as the total area of the floor, walls and roof separating the interior volume (i.e. the conditioned space) from the outside environment.

Conversion between carbon and carbon dioxide indices

0.21 To maintain consistency with the Government's Climate Change Programme, the performance targets in this Approved Document are quoted in terms of kg of carbon rather than kg of carbon dioxide, or in energy terms such as GigaJoules or MegaWatt-hours. To convert from the carbon to carbon dioxide basis multiply by the ratio of atomic weights (Carbon Dioxide 44 : Carbon 12). For example 9 tonnes per square metre per year of carbon is equivalent to {9 x (44/12)} = 33 tonnes of carbon dioxide per square metre per year.

14 IP 5/98 Metal cladding: assessing thermal performance, BRE, 1998

15 Guide to assessment of the thermal performance of aluminium curtain wall framing, CAB, 1996

16 Guide to good practice for asessing glazing frame U-values, CWCT (1998, new edition in preperation).

17 Guide to good practice for assessing heat transfer and condensation risk for a curtain wall, CWCT (1998, new edition in preperation).

18 U-value calculation procedure for light steel frame walls, BRE, to be published

19 BS EN ISO 10211-1:1996 Thermal bridges in building construction – Calculation of heat flows and surface temperatures – Part 1: General methods

20 BS EN ISO 10211-2:2001 Thermal bridges in building construction – Calculation of heat flows and surface temperatures – Part 2: Linear thermal bridges

21 BS EN 12524:2000 Building materials and products – Hygrothermal properties – Tabulated design values

22 CIBSE Guide A: Environmental design, Section A3: Thermal properties of building structures, CIBSE, 1999

Special cases

Low levels of heating

0.22 Buildings or parts of buildings with low levels of heating or no heating do not require measures to limit heat transfer through the fabric. The insulation properties of the fabric containing such spaces are chosen for operational reasons and can be regarded as reasonable provision. Low levels of heating might be no more than 25W/m^2, an example being perhaps a warehouse where heating is intended to protect goods from condensation or frost, with higher temperatures provided only around local work stations. A cold-store is an example of a building where insulation properties would be dictated by operational needs.

Low levels of use

0.23 For buildings with low hours of use, lower standards for heating and lighting systems may be appropriate, but fabric insulation standards should be no worse than the guidance in this Approved Document. Buildings used solely for worship at set times could be one example where this paragraph would be relevant.

Historic Buildings

0.24 Advice on the factors determining the character of historic buildings is set out in PPG15[23]. Specific guidance on meeting the requirements of Part L when undertaking work in historic buildings is given in Section 4 of this Approved Document.

Buildings constructed from sub-assemblies

0.25 Buildings constructed from sub-assemblies that are delivered newly made or selected from a stock are no different to any other new building and must comply with all the requirements in Schedule 1.

0.26 In some applications however, such as buildings constructed to accommodate classrooms, medical facilities, offices and storage space to meet temporary accommodation needs, reasonable external fabric provisions for the conservation of fuel and power can vary depending on the circumstances in the particular case.
For example:-

a) A building created by dismantling, transporting and re-erecting on the same premises the external fabric sub-assemblies of an existing building would normally be considered to meet the requirement;

b) A building constructed from external fabric sub-assemblies obtained from other premises or from a stock manufactured before 31 December 2001, would normally be considered to meet the requirement if the fabric thermal resistance or the prospective annual energy use will be no worse than the relevant performance standards given in the 1995 edition of Approved Document L.

0.27 Enclosed and heated or cooled links between such temporary accommodation, which may also be formed from sub-assemblies, should be insulated and made airtight to the same standards as the buildings themselves.

0.28 Where heating and lighting are to be provided in such temporary accommodation, the requirements may be satisfied in the following ways although the extent of the provisions will depend on the circumstances in the particular case:-

a) **heating and hot water systems:** providing on/off, time and temperature controls as indicated in Section 1 of this Approved Document.

b) **general and display lighting:** providing general lighting systems with lamp efficacies that are not less than those indicated in paragraph 1.48 and providing display lighting with installed efficacies not less that those indicated in paragraphs 1.50-1.52.

Mixed use development

0.29 When constructing buildings that have parts used as dwellings as well as parts for other uses, account should be taken of the guidance in Approved Document L1.

---

23 Planning and the historic environment, Planning Policy Guidance PPG 15, DoE/DNH, September 1994. (In Wales refer to Planning Guidance Wales Planning Policy First Revision 1999 and Welsh Office Circular 61/96 Planning and Historic Environment: Historic Buildings and Conservation Areas.)

Section 1: Design

General

1.1 In order to achieve energy efficiency in practice, the building and its services systems should be appropriately designed (Section 1) and constructed (Section 2). Information should also be provided such that the performance of the building in use can be assessed (Section 3). This Approved Document provides guidance on meeting the requirements at each of these important stages of procuring a building, whether it be a new building, an extension or a refurbishment project. More detailed guidance on energy efficiency measures can be found in the CIBSE Guide on Energy Efficiency in Buildings[24].

1.2 When designing building services installations, provision should be made to facilitate appropriate inspection and commissioning (see paragraphs 2.5 – 2.6).

1.3 Specific guidance for work carried out on existing buildings is given in Section 4.

1.4 In large complex buildings, it may be sensible to consider the provisions for the conservation of fuel and power separately for the different parts of the building in order to establish the measures appropriate to each part.

1.5 Where alternative building services systems are provided (e.g. dual fuel boilers, and combined heat and power or heat pump systems paralleled by standby boiler capacity), then the building should meet the requirements in each possible operating mode.

Alternative methods of showing compliance

1.6 Three methods are given for demonstrating that reasonable provision has been made for the conservation of fuel and power. These different methods offer increasing design flexibility in return for greater demands in terms of the extent of calculation required. However the overall aim is to achieve the same standard in terms of carbon emissions. The methods are:

a) an Elemental Method (paragraphs 1.7 – 1.68). This method considers the performance of each aspect of the building individually. To comply with the provisions of Part L, a minimum level of performance should be achieved in each of the elements. Some flexibility is provided for trading off between different elements of the construction, and between insulation standards and heating system performance.

b) a Whole-Building Method (paragraphs 1.69 - 1.73). This method considers the performance of the whole building. For office buildings, the heating, ventilation, air conditioning and lighting systems should be capable of being operated such that they will emit no more carbon per square metre per annum than a benchmark based on the ECON 19 data[25]. Alternative methods are also provided for schools and hospitals.

c) a Carbon Emissions Calculation Method (paragraphs 1.74 – 1.76). This method also considers the performance of the whole building, but can be applied to any building type. To comply with the provisions of Part L, the annual carbon emissions from the building should be no greater than that from a notional building that meets the compliance criteria of the Elemental Method. The carbon emissions from the proposed building and the notional building need to be estimated using an appropriate calculation tool.

Elemental Method

1.7 To show compliance following the Elemental Method, the building envelope has to provide certain minimum levels of insulation, and the various building services systems each have to meet defined minimum standards of energy efficiency as follows -

Standard U-values for construction elements

1.8 The requirement will be met if the thermal performances of the construction elements are no worse than those listed in Table 1 (as illustrated in Diagram 1). One way of achieving the U-values in Table 1 is by providing insulation of a thickness estimated from the Tables in Appendix A as illustrated in the examples. An alternative for floors is to use the data given in Appendix C.

---

24 CIBSE Guide, Energy efficiency in buildings, CIBSE, 1998

25 Energy use in offices - Energy Consumption Guide 19, DETR, 1998

Table 1 Standard U-values of construction elements

Exposed Element	U-value (W/m²K)
Pitched roof [1,2] with insulation between rafters	0.20
Pitched roof [1] with insulation between joists	0.16
Flat roof [3] or roof with integral insulation	0.25
Walls, including basement walls	0.35
Floors, including ground floors and basement floors	0.25
Windows, roof windows and personnel doors (area weighted average for the whole building), glazing in metal frames[4]	2.2
Windows, roof windows and personnel doors (area weighted average for the whole building), glazing in wood or PVC frames[4]	2.0
Rooflights [5, 6]	2.2
Vehicle access and similar large doors	0.7

Notes to Table 1:

[1] Any part of a roof having a pitch greater or equal to 70° can be considered as a wall.

[2] For the sloping parts of a room-in-the-roof constructed as a material alteration, a U-value of 0.3 W/m²K would be reasonable.

[3] Roof of pitch not exceeding 10°

[4] Display windows, shop entrance doors and similar glazing are not required to meet the standard given in this table.

[5] This standard applies only to the performance of the unit excluding any upstand. Reasonable provision would be to insulate any upstand, or otherwise isolate it from the internal environment.

[6] For the purposes of Approved Document L, a roof window may be considered as a rooflight.

Where an element is exposed to the outside via an unheated space (e.g. an unheated atrium or an underground car park), either:

a) the unheated space may be disregarded so that the element is considered as directly exposed to the outside, or

b) the U-value of the element may be calculated as the transmission heat loss coefficient through the unheated space divided by the area of the element. The transmission heat loss coefficient should be calculated as given in BS EN ISO 13789[26].

Diagram 1 Standard u-values for non-domestic buildings

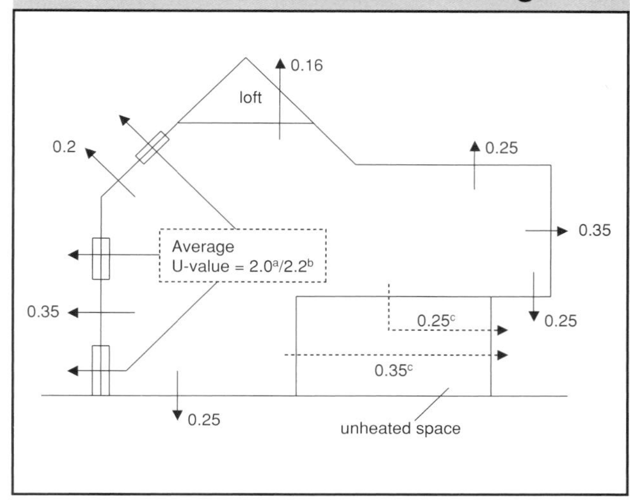

ᵃ if windows have wood or PVC frames
ᵇ if windows have metal frames
ᶜ includes the effect of the unheated space

Thermal bridging at junctions and around openings

1.9 The building fabric should be constructed so that there are no significant thermal bridges or gaps in the insulation layer(s) within the various elements of the fabric, at the joints between elements, and at the edges of elements such as those around window and door openings.

1.10 One way of demonstrating compliance would be to utilise details and practice that have been independently demonstrated as being satisfactory. For domestic style construction, a selection of such satisfactory details is given in the robust construction details publication[27].

1.11 An alternative way of meeting the requirements would be to demonstrate by calculation or by adopting robust design practices that the performance of the building is at least as good as it would be by following paragraph 1.10. BRE Information Paper IP 17/01[28] and the MRCMA Technical Report No 14[29] illustrate how this can be done.

[26] BS EN ISO 13789:1999 Thermal performance of buildings - Transmission heat loss coefficient - Calculation method

[27] Limiting thermal bridging and air leakage: Robust construction details for dwellings and similar buildings; TSO, 2001

[28] BRE IP 17/01, Assessing the effects of thermal bridging at junctions and around openings in the external elements of buildings

[29] Guidance for the design of metal cladding and roofing to comply with Approved Document L 2002 Edition: MCRMA Technical Note 14

Maximum areas of windows, doors and rooflights

1.12 Provision should be made to limit the rate of heat loss through glazed elements of the building. One way of complying would be to limit the total area of windows, doors and rooflights so that they do not exceed the values given in Table 2 - unless compensated for in some other way.

Table 2 **Maximum area of openings unless compensating measures are taken**

Building type	Windows[1] and doors as % of the area of exposed wall[2]	Rooflights as % of area of roof
Residential buildings (where people temporarily or permanently reside)	30	20
Places of assembly, offices and shops	40	20
Industrial and storage buildings	15	20
Vehicle access doors and display windows and similar glazing	As required	

Notes:
[1] For the purposes of this calculation, dormer windows in a roof may be included in the rooflight area.
[2] See paragraph 0.19, basis for calculating areas.

1.13 Care should be taken in the selection and installation of glazed systems to avoid the risk of condensation. Guidance can be obtained from BRE Report No 262[30].

Trade-off between construction elements

1.14 In order to provide greater design flexibility, the U-values of construction elements and the areas of windows, doors and rooflights may vary from the values given in Table 1 and Table 2 provided that suitable compensating measures are taken. If glazing areas are reduced from those given in Table 2, special care needs to be given to confirm that levels of daylight are adequate. Guidance on designing for daylight is contained in CIBSE LG10[31].

1.15 Compliance with the provisions of Part L would be achieved if:

a) the rate of heat loss from the proposed building does not exceed that from a notional building of the same size and shape that meets the criteria set out in Table 1 and Table 2; and

b) the U-value of any part of an element is no worse than the values given in the following Table 3.

Table 3 **Poorest U-values acceptable when trading off between construction elements**

Element	Poorest acceptable U-value(W/m^2K)
Parts of roof[1]	0.35
Parts of exposed wall or floor[1]	0.70

Notes:
[1] Whilst parts of these elements may (within the limits given in this table) have poorer U-values than those given in Table 1, it will not normally be practical to make sufficient allowances elsewhere in the design for the whole element to be built to these standards.

1.16 As further constraints in these methods however:

a) if the U-value of the floor in the proposed building is better than the performance given in Table 1 with no added insulation, the better performance standard is to be adopted for the notional building; and

b) if the area of openings in the proposed building is less than the values shown in Table 2, the average U-value of the roof, wall or floor cannot exceed the appropriate value given in Table 1 by more than 0.02 W/m^2K.

c) no more than half of the allowable rooflight area can be converted into an increased area of window and doors.

[30] Thermal insulation: avoiding risks, BR 262, 2002 Edition BRE, 2001

[31] CIBSE, Daylight and window design, LG10, CIBSE, 1999

Building air leakage standards

1.17 Buildings should be reasonably airtight to avoid unnecessary space heating and cooling demand and to enable the effective performance of ventilation systems.

1.18 Without prejudice to the need for compliance with all the requirements in Schedule 1 however, the need to provide for adequate ventilation for health (Part F) and adequate air for combustion appliances (Part J) should particularly be taken into account.

1.19 A way of meeting the requirement would be to incorporate sealing measures to achieve the performance standard given in paragraph 2.4. Some ways of achieving satisfactory airtightness include:

a) providing a reasonably continuous air barrier in contact with the insulation layer over the whole thermal envelope (including separating walls). Special care should be taken at junctions between elements and around door and window openings. For domestic type constructions, some satisfactory design details and installation practice are described in the robust details publication[25]. Guidance for the design of metal cladding and roofing systems to minimise air infiltration is given in the MCRMA Technical Report No 14[27].

b) sealing gaps around service penetrations.

c) draught-proofing external doors and windows.

Avoiding solar overheating

1.20 Buildings should be constructed such that:

a) those occupied spaces[32] that rely on natural ventilation should not overheat when subject to a moderate level of internal heat gain and

b) those spaces that incorporate mechanical ventilation or cooling do not require excessive cooling plant capacity to maintain the desired space conditions.

1.21 Ways of meeting the requirement would be through:

a) the appropriate specification of glazing, and

b) the incorporation of passive measures such as shading (detailed guidance being given in BRE Report No 364[33]), and

c) the use of exposed thermal capacity combined with night ventilation (detailed guidance being given in GIR 31[34]).

1.22 A way of achieving compliance for spaces with glazing facing only one orientation would be to limit the area of glazed opening as a percentage of the internal area of the element under consideration to the values given in Table 4.

Table 4 **Maximum allowable area of glazing**

Orientation of opening	Maximum allowable area of opening (%)
N	50
NE/NW/S	40
E/SE/W/SW	32
Horizontal	12

1.23 Alternative approaches to achieving compliance include:

a) showing that the solar heat load per unit floor area averaged between the hours of 07:30 and 17:30 would not be greater than $25W/m^2$ if the building were to be subject to the solar irradiances for the particular location for the month of July that were not exceeded on more than 2.5% of occasions during the period 1976 – 1995. The procedure given in Appendix H can be used to do this.

b) showing by detailed calculation procedures such as those described in chapter 5 of CIBSE Guide A[35], that in the absence of mechanical cooling or mechanical ventilation, the space will not overheat when subjected to an internal gain of $10 W/m^2$.

Heating systems

1.24 Without prejudice to the need for compliance with all the requirements in Schedule 1, the need to provide for adequate ventilation for health (Part F) and adequate air for combustion appliances (Part J) should particularly be taken into account when making provisions for combustion systems such as boiler plant and direct-fired gas heaters.

Carbon intensities of heating plant

General

1.25 Heating plant should be reasonably efficient. For heating plant serving hot water and steam heating systems, electric heating and heat pumps (irrespective of the form of heat distribution), a way of complying with the requirement would be to show that:

[32] This guidance is not applicable to non-occupied spaces such as stacks and atria intended to drive natural ventilation via buoyancy.

[33] Solar shading of buildings, BR 364, CRC Ltd, 1999.

[34] Energy Efficiency Best Practice programme, Avoiding or minimising the use of air-conditioning, GIR 31, TSO, 1995.

[35] Environmental design, Guide A, CIBSE, 1999.

a) the carbon intensity of the heat generating equipment at the maximum heat output of the heating system[36] is not greater than the value shown in Table 5 column (a) and

b) the carbon intensity of the heat generating equipment when the system is producing 30% of the maximum heat output of the heating system is not greater than the value shown in Table 5 column (b).

Table 5 Maximum allowable carbon intensities of heating systems

Fuel	Maximum carbon intensities (kgC/kWh) at stated % of the maximum heat output of the heating system. (a) at maximum heat output	(b) At 30% of maximum heat output
Natural gas	0.068	0.065
Other fuels	0.091	0.088

1.26 In some applications electric resistance heating might be appropriate. In such cases the designer will need to improve envelope insulation standards to trade-off against the higher carbon intensity of such forms of heating (see paragraph 1.32).

1.27 The carbon intensity of the heating plant is based on the carbon emitted per useful kWh of heat output. Where there are multiple pieces of heat generating equipment, the carbon intensity of the heating plant is the rating weighted average of the individual elements. This definition of carbon intensity is applicable to boilers, heat pump systems and electrical heating, and is given by:

$$\varepsilon_c = \frac{1}{\sum R} \sum \frac{R.C_f}{\eta_t} \qquad (1)$$

where:

ε_c = the carbon intensity of the heating system (kgC/kWh of useful heat).

R = the rated output of an individual element of heat raising plant (kW).

η_t = the gross thermal efficiency of that element of heat raising plant (kWh of heat per kWh of delivered fuel). For most practical cases, the efficiency may be taken as the full load efficiency for that element but, where appropriate, a part load efficiency based on manufacturer's certified data may be used as an alternative.

C_f = the carbon emission factor of the fuel supplying that element of heat raising plant (kg of carbon emitted per kWh of delivered fuel consumed) (Table 6).

Table 6 Carbon emission factors

Delivered fuel	Carbon emission factor (kgC/kWh)
Natural gas	0.053
LPG	0.068
Biogas	0
Oil [1]	0.074
Coal	0.086
Biomass	0
Electricity [2]	0.113
Waste heat [3]	0

Notes:
[1] This value can be used for all grades of fuel oil.
[2] This is the estimated average figure for grid-supplied electricity for the period 2000 – 2005. If there is on-site generation of electricity from photovoltaic panels or wind power, this could be accounted for using the carbon emissions calculation method (see paragraph 1.74 et seq). Short term energy supply arrangements such as "green tariffs" are not appropriate ways of complying with the requirements.
[3] This includes waste heat from industrial processes and from power stations rated at more than 10MW electrical output and with a power efficiency > 35%

Calculating the carbon intensity of CHP systems

1.28 Where a combined heat and power system (CHP) is proposed, the carbon intensity of the CHP (ε_{chp}) can take account of the benefit of the on-site generation in reducing emissions from power stations feeding the national grid. The adjustment can be made using equation (2). This adjusted carbon intensity can then be used in equation (1) to determine the carbon intensity of the overall heating system at 100% and 30% of heating system output.

$$\varepsilon_{chp} = \frac{C_f}{\eta_t} - \frac{C_{displaced}}{HPR} \qquad (2)$$

where:

η_t is the gross thermal efficiency of the CHP engine (kWh of useful heat per kWh of fuel burned);

HPR is the heat to power ratio (kWh of useful heat produced per kWh of electricity output). This is equivalent to the ratio of the thermal and power efficiencies of the CHP unit;

$C_{displaced}$ is the carbon emission factor for grid-supplied electricity displaced by the CHP (kg/kWh). This should be taken as the factor for new generating capacity that might otherwise be built if the CHP had not been provided, i.e. the intensity of a new generation gas-fired station: 0.123 kgC/kWh.

[36] The maximum heat output of the heating system occurs when all the items of heat generating equipment capable of operating simultaneously are producing heat at their rated output.

1.29 Where the CHP has no facility for heat dumping, the gross thermal efficiency is simply the CHP heat output divided by the energy content of the fuel burned. Where the CHP includes facilities for heat dumping, the gross thermal efficiency should be based on an estimate of the useful heat supplied to the building, i.e. the heat output from the CHP minus the heat dumped. Certification under the CHPQA[37] scheme would be a way of showing that the gross thermal efficiency has been estimated in a satisfactory way.

Calculating the carbon intensity of community heating

1.30 When calculating the carbon intensity of the heat supplied to the building by a community heating system, account should be taken of the performance of the whole system, (i.e. the distribution circuits, and all the heat generating plant including any CHP or waste heat recovery) and the carbon emission factors of the different fuels. Certification under the CHPQA[37] scheme would be a way of showing that the thermal and power efficiencies have been estimated in a satisfactory way.

Other methods of heating

1.31 In buildings such as factories, warehouses and workshops it can be more efficient to provide local warm air or radiant heating systems. GPG 303[38] provides guidance on the application of such systems.

Trade - off between construction elements and heating system efficiency

1.32 In order to allow greater design flexibility, there can be a trade-off (in either direction) between the average U-value of the envelope and the carbon intensity of the heating system provided that the rate of carbon emissions is unchanged. A way of complying would be to adjust the average U-value of the building fabric such that it is no worse than the value determined from the following equation.

$$U_{req} = U_{ref} \cdot \frac{\varepsilon_{ref}}{\varepsilon_{act}} \qquad (3)$$

where:

U_{req} = the required average U-value.

U_{ref} = the average U-value of the building constructed to the elemental standards of Table 1.

ε_{ref} = the carbon intensity of the reference heating system at an output of 30% of the installed design capacity. This should be taken from Table 5 column (b) for the fuel type used in the actual heating system.

ε_{act} = the carbon intensity of the actual heating system at an output of 30% of installed design capacity.

Space heating controls

1.33 The building should be provided with zone, timing and temperature controls such that each functional area is maintained at the required temperature only during the period when it is occupied. Additional controls may be provided to allow heating during extended unusual occupation hours and to provide for sufficient heating to prevent condensation or frost damage when the heating system would otherwise be switched off.

1.34 Ways of meeting the requirement include:

a) in buildings with a heating system maximum output not exceeding 100 kW, to follow the guidance in GPG 132[39].

b) in larger or more complex buildings, to follow the guidance contained in CIBSE Guide H[40]. Certification by a competent person that the provisions meet the requirements may be accepted by building control bodies.

Hot water systems and their control

1.35 Hot water should be provided safely, making efficient use of energy and thereby minimising carbon emissions. Ways of achieving the requirement include -

a) avoiding over-sizing of hot water storage systems.

b) avoiding low-load operation of heat raising plant.

c) avoiding the use of grid-supplied electric water heating except where hot water demand is low.

d) providing solar water heating.

e) minimising the length of circulation loops.

f) minimising the length and diameter of dead legs.

1.36 A way of satisfying the requirements for conventional hot water storage systems would be to provide controls that shut off heating when the required water temperature is achieved. The supply of heat should also be shut off during those periods when hot water is not required.

---

37 Quality Assurance for Combined Heat and Power, CHPQA Standard, Issue 1, DETR, November 2000.

38 GPG 303, 2000: The designer's guide to energy-efficient buildings for industry, Energy Efficiency Best practice programme, BRECSU.

39 GPG 132, 2001: Energy Efficiency Best Practice Programme, Heating controls in small commercial and multi-residential buildings, BESCSU.

40 CIBSE Guide H, 2000: Building Control Systems.

1.37 Ways of meeting the requirement include:-

a) in small buildings, following the guidance in GPG 132[39].

b) in larger, more complex buildings or for alternative systems (e.g. solar hot water heating), following the guidance contained in CIBSE Guide H[40]. Certification by a competent person that the provisions meet the requirements may be accepted by building control bodies.

Insulation of pipes, ducts and vessels

Limit of application:

1.38 This section only applies to pipework, ductwork and vessels for the provision of space heating, space cooling (including chilled water and refrigerant pipework) and hot water supply for normal occupation. Pipework, ductwork and vessels for process use are outside the scope of the Building Regulations.

Meeting the requirement

1.39 A way of meeting the requirement would be to apply insulation to the standards required in BS 5422[41] to all pipework, ductwork and storage vessels. The requirement for storage vessels should be taken as that given in BS 5422 for flat surfaces.

1.40 Insulation would not be necessary for compliance with Part L if the heat flow through the walls of the pipe, duct or vessel is always useful in conditioning the surrounding space when fluid is flowing through the pipe or duct, or is being stored in the vessel in question. However, it may be prudent to provide it to facilitate control stability.

Lighting efficiency standards

1.41 Lighting systems should be reasonably efficient and make effective use of daylight where appropriate.

1.42 For the purposes of Approved Document L circuit-watts means the power consumed in lighting circuits by lamps and their associated control gear and power factor correction equipment.

General lighting efficacy in office, industrial and storage buildings

1.43 Electric lighting systems serving these buildings should be provided with reasonably efficient lamp/luminaire combinations. A way of meeting the requirements would be to provide lighting with an initial efficacy averaged over the whole building of not less than 40 luminaire-lumens/circuit-watt. This allows considerable design flexibility to vary the light output ratio of the luminaire, the luminous efficacy of the lamp or the efficiency of the control gear.

1.44 The average luminaire-lumens/circuit-watt is calculated by

$$\eta_{lum} = \frac{1}{P} \cdot \sum \frac{LOR.\phi_{lamp}}{C_L} \qquad (4)$$

where η_{lum} = the luminaire efficacy (luminaire-lumens/circuit-watt);

LOR = the light output ratio of the luminaire, which means the ratio of the total light output of a luminaire under stated practical conditions to that of the lamp or lamps contained in the luminaire under reference conditions;

ϕ_{lamp} = the sum of the average initial (100 hour) lumen output of all the lamp(s) in the luminaire;

P = the total circuit watts for all the luminaires;

C_L = the factor applicable when controls reduce the output of the luminaire when electric light is not required. The values of C_L are given in Table 7.

Table 7 **Luminaire control factors**

Control function	C_L
a) The luminaire is in a daylit space (see paragraph 1.45), and its light output is controlled by • A photoelectric switching or dimming control, with or without manual override, or • Local manual switching (see paragraph 1.57a)	0.80
b) The luminaire is in a space that is likely to be unoccupied for a significant proportion of working hours and where a sensor switches off the luminaire in the absence of occupants but switching on is done manually	0.80
c) Circumstances a) and b) above combined.	0.75
d) None of the above.	1.00

1.45 For the purposes of this Approved Document, a daylit space is defined as any space within 6m of a window wall, provided that the glazing area is at least 20% of the internal area of the window wall. Alternatively it can be roof-lit, with a glazing area at least 10% of the floor area. The normal light transmittance of the glazing should be at least 70%, or, if the light transmittance is reduced below 70%, the glazing area could be increased proportionately, but subject to the considerations given in paragraphs 1.12 and 1.20 – 1.23.

41 BS 5422: 2001 Methods for specifying thermal insulation materials on pipes, ductwork and equipment in the temperature range -40°C to +700°C.

1.46 This guidance need not be applied in respect of a maximum of 500 W of installed lighting in the building, thereby allowing flexibility for the use of feature lighting etc.

1.47 Appendix F gives examples that show how the luminaire efficacy requirement can be met either by selection of appropriate lamps and luminaires or by calculation.

General lighting efficacy in all other building types

1.48 For electric lighting systems serving other building types, it may be appropriate to provide luminaires for which photometric data is not available and/or are lower powered and use less efficient lamps. For such spaces, the requirements would be met if the installed lighting capacity has an initial (100 hour) lamp plus ballast efficacy of not less than 50 lamp-lumens per circuit-watt. A way of achieving this would be to provide at least 95% of the installed lighting capacity using lamps with circuit efficacies no worse than those in Table 8.

Table 8 Light sources meeting the criteria for general lighting

Light source	Types and ratings
High pressure Sodium	All types and ratings
Metal halide	All types and ratings
Induction lighting	All types and ratings
Tubular fluorescent	26mm diameter (T8) lamps, and 16mm diameter (T5) lamps rated above 11W, provided with high efficiency control gear. 38mm diameter (T12) linear fluorescent lamps 2400mm in length
Compact fluorescent	All ratings above 11W
Other	Any type and rating with an efficacy greater than 50 lumens per circuit Watt.

1.49 For the purposes of Approved Document L, high efficiency control gear means low loss or high frequency control gear that has a power consumption (including the starter component) not exceeding that shown in Table 9[42].

Table 9 Maximum power consumption of high efficiency control gear

Nominal lamp rating (Watts)	Control gear power consumption (Watts)
Less than or equal to 15	6
Greater than 15, Not more than 50	8
Greater than 50, Not more than 70	9
Greater than 70, Not more than 100	12
Greater than 100	15

Display lighting in all buildings

1.50 For the purposes of Approved Document L, display lighting means lighting intended to highlight displays of exhibits or merchandise, or lighting used in spaces for public entertainment such as dance halls, auditoria, conference halls and cinemas.

1.51 Because of the special requirements of display lighting, it may be necessary to accept lower energy performance standards for display lighting. Reasonable provision should nevertheless be made and a way of complying would be to demonstrate that the installed capacity of display lighting averaged over the building has an initial (100 hour) efficacy of not less than 15 lamp-lumens per circuit-watt. In calculating this efficacy, the power consumed by any transformers or ballasts should be taken into account.

1.52 As an alternative, it would be acceptable if at least 95% of the installed display lighting capacity in circuit-Watts comprises lighting fittings incorporating lamps that have circuit efficacies no worse than those in Table 10.

Table 10 Light sources meeting the criteria for display lighting

Light source	Types and ratings
High pressure Sodium	All types and ratings
Metal halide	All types and ratings
Tungsten halogen	All types and ratings
Compact and tubular fluorescent	All types and ratings
Other	Any type and rating with an efficacy greater than 15 lumens per circuit Watt.

[42] The values in the table are in line with European Directive 2000/55/EC 18 September 2000 : On energy efficiency requirements for ballasts for fluorescent lighting

Emergency escape lighting and specialist process lighting

1.53 For the purposes of Approved Document L, the following definitions apply:

a) Emergency escape lighting means that part of emergency lighting that provides illumination for the safety of people leaving an area or attempting to terminate a dangerous process before leaving an area.

b) Specialist process lighting means lighting intended to illuminate specialist tasks within a space, rather than the space itself. It could include theatre spotlights, projection equipment, lighting in TV and photographic studios, medical lighting in operating theatres and doctors' and dentists' surgeries, illuminated signs, coloured or stroboscopic lighting, and art objects with integral lighting such as sculptures, decorative fountains and chandeliers.

1.54 Emergency escape lighting and specialist process lighting are not subject to the requirements of Part L.

Lighting controls

1.55 Where it is practical, the aim of lighting controls should be to encourage the maximum use of daylight and to avoid unnecessary lighting during the times when spaces are unoccupied. However, the operation of automatically switched lighting systems should not endanger the passage of building occupants. Guidance on the appropriate use of lighting controls is given in BRE IP 2/99[43].

Controls in offices and storage buildings

1.56 A way of meeting the requirement would be the provision of local switches in easily accessible positions within each working area or at boundaries between working areas and general circulation routes. For the purposes of Approved Document L2, reference to switch includes dimmer switches and switching includes dimming. As a general rule, dimming should be effected by reducing rather than diverting the energy supply.

1.57 The distance on plan from any local switch to the luminaire it controls should generally be not more than eight metres, or three times the height of the light fitting above the floor if this is greater. Local switching can be supplemented by other controls such as time switching and photo-electric switches where appropriate. Local switches could include:

a) switches that are operated by the deliberate action of the occupants either manually or by remote control. Manual switches include rocker switches, push buttons and pull cords. Remote control switches include infra red transmitter, sonic, ultrasonic and telephone handset controls.

b) automatic switching systems which switch the lighting off when they sense the absence of occupants.

Controls in buildings other than offices and storage buildings

1.58 A way of meeting the requirement would be to provide one or more of the following types of control system arranged to maximise the beneficial use of daylight as appropriate:

a) local switching as described in paragraph 1.57;

b) time switching, for example in major operational areas which have clear timetables of occupation;

c) photo-electric switching.

Controls for display lighting (all building types)

1.59 A way of meeting the requirement would be to connect display lighting in dedicated circuits that can be switched off at times when people will not be inspecting exhibits or merchandise or being entertained. In a retail store, for example, this could include timers that switch the display lighting off outside store opening hours, except for displays designed to be viewed from outside the building through display windows.

Air-conditioning and mechanical ventilation (ACMV)

1.60 For the purposes of Approved Document L, the following definitions apply:

a) mechanical ventilation is used to describe systems that use fans to supply outdoor air and/or extract indoor air to meet ventilation requirements. Systems may be extensive and can include air filtration, air handling units and heat reclamation, but they do not provide any active cooling from refrigeration equipment. The definition would not apply to a naturally ventilated building, which makes use of individual wall or window mounted extract fans to improve the ventilation of a small number of rooms.

b) air conditioning is used to describe any system where refrigeration is included to provide cooling for the comfort of building occupants. Air conditioning can be provided from stand-alone refrigeration equipment in the cooled space, from centralised or partly centralised equipment, and from systems that combine the cooling function with mechanical ventilation.

---

[43] Photoelectric control of lighting: design, set-up and installation issues, BRE Information Paper IP 2/99

c) treated areas; these are the floor areas of spaces that are served by the mechanical ventilation or air conditioning system in the context and should be established by measuring between the internal faces of the surrounding walls. Spaces that are not served by these systems such as plant rooms, service ducts, lift-wells etc. should be excluded.

d) process requirements; in office buildings process requirements can be taken to include any significant area in which an activity takes place which is not typical of the office sector, and where the resulting need for heating, ventilation or air conditioning is significantly different to that of ordinary commercial offices. When assessing the performance of air conditioning or mechanical ventilation systems, areas which are treated because of process requirements should be excluded from the treated area, together with the plant capacity, or proportion of the plant capacity, that is provided to service those areas. Activities and areas in office buildings considered to represent process requirements would thus include:

- Staff restaurants and kitchens;
- Large, dedicated, conference rooms;
- Sports facilities;
- Dedicated computer or communications rooms.

e) In the following text, air conditioning and/or mechanical ventilation systems as defined above are collectively described as ACMV.

1.61 Buildings with ACMV should be designed and constructed such that:

a) the form and fabric of the building do not result in a requirement for excessive installed capacity of ACMV equipment. In particular, the suitable specification of glazing ratios and solar shading are an important way to limit cooling requirements (see paragraphs 1.20 – 1.23).

b) components such as fans, pumps and refrigeration equipment are reasonably efficient and appropriately sized to have no more capacity for demand and standby than is necessary for the task.

c) suitable facilities are provided to manage, control and monitor the operation of the equipment and the systems.

CPR method for office buildings with ACMV

1.62 In the case of an office development, one way of achieving compliance (if there are no innovative building or building services provisions) is to show that the Carbon Performance Rating (CPR) is satisfactory. Where there are innovative features in the design the carbon emissions calculation method, or another acceptable alternative would be more appropriate.

1.63 The CPR is a rating based on standardised occupancy patterns that relates the performance of the proposed building to a benchmark based on the measured consumption data contained in ECON19[23]. The rating of the proposed building is calculated from the rated input power of the installed equipment as this combines the effect of load reduction by good envelope design and energy efficient system design into a single parameter. The detail of the CPR calculation is contained in Appendix G. If there are any areas in the building with significant process loads (e.g. a major computer suite), such areas and their associated plant capacity should be excluded from the calculation of the CPR. However, in order to facilitate comparison with operational performance, such discounted loads should be separately metered (see paragraph 3.6d)). For an illustration of this calculation method, see Appendix G.

1.64 For new ACMV systems, compliance would be achieved if the CPR is no greater than the values shown in Table 11.

Table 11 **Maximum allowable Carbon Performance Ratings**

	Maximum CPR (kgC/m²/year) for a new system installed in:	
System type	**a) a new building**	**b) an existing building**
Air conditioning	10.3	11.35
Mechanical ventilation	6.5	7.35

If both new air conditioning and mechanical ventilation systems are to be installed in a building, the system types and their treated areas should be dealt with separately and the appropriate CPR achieved for each.

1.65 For a building that already contains an ACMV system and substantial alteration is being made to that existing ACMV system, compliance would be achieved if the CPR is improved (i.e. reduced) by 10% as a result of the work, or does not exceed the values in Table 11 column b), whichever is the least demanding.

1.66 When the work solely comprises replacement of existing equipment, the product of the installed capacity per unit treated area (P) and the control management factor (F) should:-

a) be reduced by at least 10%, OR

b) meet a level of performance equivalent to the component benchmarks given in CIBSE TM22[44] (i.e. the product of service provision, efficiency and control factor), whichever is the least demanding.

Methods for other buildings with ACMV

1.67 For other buildings, it is only possible at present to define an overall performance requirement for the mechanical ventilation systems (whether or not the air being supplied and/or extracted is heated or cooled). In such cases, the requirement can be met if the specific fan power (SFP) is less than the values given in the following sub-paragraphs. The specific fan power is the sum of the design total circuit-watts, including all losses through switchgear and controls such as inverters, of all fans that supply air and exhaust it back to outdoors (i.e., the sum of supply and extract fans), divided by the design ventilation rate through the building.

a) for ACMV systems in new buildings, the SFP should be no greater than 2.0 W/litre/second.

b) for new ACMV systems in refurbished buildings, or where an existing ACMV system in an existing building is being substantially altered, the SFP should be no greater than 3.0 W/litre/second.

These SFP values are appropriate to typical spaces ventilated for human occupancy. Where specialist processes or higher than normal external pollution levels require increased levels of filtration or air cleaning, higher SFPs may be appropriate.

1.68 Mechanical ventilation systems should be reasonably efficient at part load. One way to achieve this would be to provide efficient variable flow control systems incorporating for instance variable speed drives or variable pitch axial fans. More detailed guidance on these measures is given in GIR 41[45].

Whole-building Method

1.69 To show compliance following the Whole-building Method, the carbon emissions or primary energy consumption at the level of the complete building have to be reasonable for the purpose of the conservation of fuel and power. This approach allows much more design flexibility than the Elemental method.

Office buildings

1.70 The Whole-Office Carbon Performance Rating method is a development of the CPR described in paragraphs 1.62 and 1.63. In this compliance route, the rating is expanded to cover lighting and space heating as explained in detail in BRE Digest No 457[46].

1.71 The requirement would be met if:

a) the whole-office CPR is no greater than the values shown in Table 12 AND

b) the envelope meets the requirements of paragraphs 1.9-1.11, 1.17-1.19 and Table 3.

Table 12 **Maximum whole-office CPR**

	Maximum allowable CPR ($kgC/m^2/year$)	
Building type	**New office**	**Refurbished office**
Naturally ventilated	7.1	7.8
Mechanically ventilated	10.0	11.0
Air-conditioned	18.5	20.4

Schools

1.72 For schools, a way of complying with the requirements would be to show that the proposed building conforms with the DfEE guidance note[47].

Hospitals

1.73 For hospitals, a way of complying with the requirements would be to show that the proposed building conforms with the NHS Estates guides[48].

---

44 CIBSE, Energy Assessment and Reporting Methodology: Office Assessment Method, TM22, CIBSE, 1999.

45 Variable flow control, General Information Report 41, Energy Efficiency Best Practice programme, 1996.

46 BRE Digest No 457: "The Carbon Performance Rating for offices".

47 DfEE, Guidelines for environmental design in schools, Building Bulletin 87, TSO, 1997.

48 NHS Estates: Achieving energy efficiency in new hospitals, TSO, 1994.

Carbon Emissions Calculation Method

1.74 To show compliance using the Carbon Emissions Calculation Method, the calculated annual carbon emissions of the proposed building should be no greater than those from a notional building of the same size and shape designed to comply with the Elemental Method. This approach allows more flexible design of the building, taking advantage of any valid energy conservation measure and taking account of useful solar and internal heat gains.

1.75 The following constraints should however be applied:-

a) when establishing the parameters of the notional building, the constraint on floor U-value in paragraph 1.16a) should be applied, and

b) the proposed building fabric and air leakage performances should be no worse than those given in Table 3 and paragraphs 1.9-1.11 and 1.17-1.19 respectively.

1.76 The calculations should be carried out using an acceptable method. The method may be acceptable to building control bodies if:-

a) it has been approved by a relevant authority responsible for issuing professional guidance, or

b) it has been accepted by the organisation responsible for the work as having satisfied their in-house quality assurance procedures. This could be demonstrated by submitting with the calculations a completed copy of Appendix B (Checklist for choosing BEEM software) of AM11[49], showing that the software used is appropriate for the purpose to which it has been applied.

Conservatories, atria and similar sun-spaces

1.77 For the purposes of section 1 of Approved Document L2, sun-space (which includes conservatories and atria) means a building or part of a building having not less than three-quarters of the area of its roof and not less than half the area of its external walls (if any) made of translucent material.

1.78 When a sun-space is attached to and built as part of a new building:

a) where there is no separation between the sun-space and the building, the sun-space should be treated as an integral part of the building;

b) where there is separation between the sun-space and the building, energy savings can be achieved if the sun-space is not heated or mechanically cooled. If fixed heating or mechanical cooling installations are proposed, however, they should have their own separate temperature and on/off controls.

1.79 When a sun-space is attached to an existing building and an opening is enlarged or newly created as a material alteration, reasonable provision should be made to enable the heat loss from, or the summer solar heat gain to, the building to be limited. Ways of meeting the requirement would be:

a) to retain the existing separation where the opening is not to be enlarged; or

b) to provide separation as, or equivalent to, windows and doors having the average U-value given in Table 1.

1.80 For the purposes of this Approved Document, separation between a building and a sun-space means:

a) separating walls and floors that are insulated to at least the same degree as the exposed walls and floors;

b) separating windows and doors with the same U-value and draught-proofing provisions as the exposed windows and doors elsewhere in the building.

1.81 Attention is drawn to the safety requirements of Part N of the Building Regulations regarding glazing materials and protection.

---

[49] CIBSE AM11, 1998: Building Energy and Environmental Modelling.

SECTION 2: Construction

Building Fabric

Continuity of insulation

2.1 To avoid excessive thermal bridging, appropriate design details and fixings should be used (see paragraph 1.9). Responsibility for achieving compliance with the requirements of Part L rests with the person carrying out the work. In the case of new buildings, that "person" will usually be, e.g., a developer or main contractor who has carried out the work subject to Part L, directly or by engaging a subcontractor. The person responsible for achieving compliance should (if suitably qualified) provide a certificate or declaration that the provisions meet the requirements of Part L2(a); or they should obtain a certificate or declaration to that effect from a suitably qualified person. Such certificates or declarations would state:

a) that appropriate design details and building techniques have been used and that the work has been carried out in ways that can be expected to achieve reasonable conformity with the specifications that have been approved for the purposes of compliance with Part L2; or

b) that infra-red thermography inspections have shown that the insulation is reasonably continuous over the whole visible envelope. BRE Report 176[50] gives guidance on the use of thermography for building surveys.

Airtightness

2.2 Air barriers should be installed to minimise air infiltration through the building fabric (see paragraph 1.19). In this case too, certificates or declarations should be provided or obtained by the person carrying out the work, stating:

a) for buildings of any size, that the results of air leakage tests carried out accordance with CIBSE TM 23[51] are satisfactory; or

b) alternatively for buildings of less than $1000m^2$ gross floor area, that appropriate design details and building techniques have been used, and that the work has been carried out in ways that can be expected to achieve reasonable conformity with the specifications that have been approved for the purposes of compliance with Part L2.

Certificates and Testing

2.3 Certificates or declarations such as those mentioned in paragraphs 2.1 and 2.2 may be accepted by building control bodies as evidence of compliance. The building control body will, however, wish to establish, in advance of the work, that the person who will be giving the certificates or declarations is suitably qualified.

2.4 If using the CIBSE TM 23 pressure test procedures as the means of showing compliance:-

a) With effect from 1 October 2003, reasonable provision would be test results showing that the air permeability (see paragraph 0.20) does not exceed $10m^3/h/m^2$ at an applied pressure difference of 50 Pa.

b) In the period up to and 30 September 2003, reasonable provision in the event that initial test results are unsatisfactory would be the results of further tests carried out after appropriate remedial work showing:-

i) an improvement of 75% of the difference between the initial test result and the target standard of 10 $m^3/h/m^2$ at 50 Pa; OR, if less demanding

ii) a performance no worse than 11.5 $m^3/h/m^2$ at 50 Pa.

Inspection and Commissioning of the Building Services Systems

2.5 In Part L2, in the context of building services systems, 'providing' and 'making provision' should be taken as including, where relevant, inspection and commissioning with meanings as described below:

a) Inspection of building services systems means establishing at completion of installation that the specified and approved provisions for efficient operation have been put in place.

b) Commissioning means the advancement of these systems from the state of static completion to working order to the specifications relevant to achieving compliance with Part L, without prejudice to the need to comply with health and safety requirements. For each system it includes setting-to-work, regulation (that is testing and adjusting repetitively) to achieve the specified performance, the calibration, setting up and testing of the associated automatic control systems, and recording of the system settings and the performance test results that have been accepted as satisfactory.

2.6 As noted in paragraph 2.1, responsibilty for achieving compliance with the requirements of Part L rests with the person carrying out the work. In the case of building services systems, that "person" may be, e.g., a developer or main

---

[50] A practical guide to infra-red thermography for building surveys, BRE report 176, BRE, 1991.

[51] CIBSE, Testing buildings for air leakage, TM 23, CIBSE, 2000.

contractor who has carried out the work directly, or by engaging a subcontractor to carry it out; or it may be a specialist firm directly engaged by a client. The person responsible for achieving compliance should provide a report, or obtain one from a suitably qualified person, that indicates the inspection and commissioning activities necessary to establish that the work complies with Part L have been completed to a reasonable standard. Such reports should include:

a) a commissioning plan that shows that every system has been inspected and commissioned in an appropriate sequence. A way of demonstrating compliance would be to follow the guidance in the CIBSE Commissioning Codes and BSRIA Commissioning Guides[52].

b) the results of the tests confirming the performance is reasonably in accordance with the approved designs including written commentaries where excursions are proposed to be accepted.

2.7 Such reports may be accepted by building control bodies as evidence of compliance. The building control body will, however, wish to establish, in advance of the work, that any person who will be providing such a report is suitably qualified.

---

[52] The Commissioning Specialists Association Technical Memorandum 1 Standard specification for the commissioning of mechanical engineering services installations for buildings (1999), provides a standard specification for the commissioning of mechanical engineering services installations and also gives guidance on managing the process.

SECTION 3: Providing information

Building log-book

3.1 The owner and/or occupier of the building should be provided with a log-book giving details of the installed building services plant and controls, their method of operation and maintenance, and other details that collectively enable energy consumption to be monitored and controlled. The information should be provided in summary form, suitable for day-to-day use. This summary could draw on or refer to information available as part of other documentation, such as the Operation and Maintenance Manuals and the Health and Safety file.

3.2 The details to be provided could include:

a) a description of the whole building, its intended use and design philosophy and the intended purpose of the individual building services systems;

b) a schedule of the floor areas of each of the building zones categorised by environmental servicing type (e.g. air-conditioned, naturally ventilated);

c) the location of the relevant plant and equipment, including simplified schematic diagrams;

d) the installed capacities (input power and output rating) of the services plant;

e) simple descriptions of the operational and control strategies of the energy consuming services in the building;

f) a copy of the report confirming that the building services equipment has been satisfactorily commissioned (see paragraph 2.6(b);

g) operating and maintenance instructions that include provisions enabling the specified performance to be sustained during occupation;

h) a schedule of the building's energy supply meters and sub-meters, indicating for each meter, the fuel type, its location, identification and description, and instructions on their use. The instructions should indicate how the energy performance of the building (or each separate tenancy in the building where appropriate) can be calculated from the individual metered energy readings to facilitate comparison with published benchmarks (see paragraphs G6 to G9 in Appendix G). Guidance on appropriate metering strategies is given, starting at paragraph 3.3 below;

i) for systems serving an office floor area greater than 200 m^2, a design assessment of the building services systems' carbon emissions and the comparable performance benchmark (see paragraph G4 in Appendix G);

j) the measured air permeability of the building (see paragraph 2.4).

Installation of energy meters

3.3 To enable owners or occupiers to measure their actual energy consumption, the building engineering services should be provided with sufficient energy meters and sub-meters. The owners or occupiers should also be provided with sufficient instructions including an overall metering strategy that show how to attribute energy consumptions to end uses and how the meter readings can be used to compare operating performance with published benchmarks (see paragraph 3.2.h). GIL 65[53] provides guidance on developing metering strategies.

3.4 Reasonable provision would be to enable at least 90% of the estimated annual energy consumption of each fuel to be accounted for. Allocation of energy consumption to the various end uses can be achieved using the following techniques:-

a) direct metering;

b) measuring the run-hours of a piece of equipment that operates at a constant known load;

c) estimating the energy consumption, e.g. from metered water consumption for HWS, the known water supply and delivery temperatures and the known efficiency of the water heater;

d) estimating consumption by difference, e.g. measuring the total consumption of gas, and estimating the gas used for catering by deducting the measured gas consumption for heating and hot water;

e) estimating non-constant small power loads using the procedure outlined in Chapter 11 of the CIBSE Energy Efficiency Guide[54].

---

53 GIL 65, 2001: Sub metering new build non-domestic buildings, BRECSU.

54 CIBSE Energy Efficiency Guide, 1998: Chapter 11, General electric power.

3.5 Reasonable provision of meters would be to install incoming meters in every building greater than 500m^2 gross floor area (including separate buildings on multi-building sites). This would include:

a) individual meters to directly measure the total electricity, gas, oil and LPG consumed within the building;

b) a heat meter capable of directly measuring the total heating and/or cooling energy supplied to the building by a district heating or cooling scheme.

3.6 Reasonable provision of sub-metering would be to provide additional meters such that the following consumptions can be directly measured or reliably estimated (see paragraph 3.4).

a) electricity, natural gas, oil and LPG provided to each separately tenanted area that is greater than 500m^2.

b) energy consumed by plant items with input powers greater or equal to that shown in Table 13.

c) any heating or cooling supplied to separately tenanted spaces. For larger tenancies, such as those greater than 2500m^2, direct metering of the heating and cooling may be appropriate, but for smaller tenanted areas, the heating and cooling end uses can be apportioned on an area basis.

d) any process load (see paragraph 1.60d)) that is to be discounted from the building's energy consumption when comparing measured consumption against published benchmarks.

Table 13 **Size of plant for which separate metering would be reasonable.**

Plant item	Rated input power (kW)
Boiler installations comprising one or more boilers or CHP plant feeding a common distribution circuit.	50
Chiller installations comprising one or more chiller units feeding a common distribution circuit	20
Electric humidifiers	10
Motor control centres providing power to fans and pumps	10
Final electrical distribution boards	50

SECTION 4: Work on existing buildings

Replacement of a controlled service or fitting

4.1 "Controlled service or fitting" is defined in Regulation 2(1) of the Building Regulations 2000 (as amended by the Building (Amendment) Regulations 2001) as follows:

Controlled service or fitting means a service or fitting in relation to which part G, H, J or L of Schedule 1 imposes a requirement.

4.2 Building work is defined in Regulation 3(1) to include *the provision or extension of a controlled service or fitting in or in connection with a building.*

4.3 Reasonable provision where undertaking replacement work on controlled services or fittings (whether replacing with new but identical equipment or with different equipment and whether the work is solely in connection with controlled services or includes work on them) depends on the circumstances in the particular case and would also need to take account of historic value (see paragraph 4.10). Possible ways of meeting the requirements include the following:

a) **Windows, doors and rooflights.** When these elements are to be replaced, provide units that meet the requirements for new buildings or that provide a centre-pane U-value no worse than 1.2 W/m²K (the requirement does not apply to repair work on parts of these elements, such as replacing broken glass or sealed double-glazing units or replacing rotten framing members). The replacement work should comply with the requirements of Part L and (unless only non-glazed fittings are involved) Part N. In addition the building should not have a worse level of compliance, after the work, with other applicable Parts of Schedule 1. These may include Parts B, F and J.

b) **Heating systems.** Where heating systems are to be substantially replaced, providing a new heating system and controls as if they are new installations. In lesser work, make reasonable provision for insulation, zoning, timing, temperature and interlock controls. Without prejudice to the need for compliance with all the requirements in Schedule 1, the need to comply with the requirements of Parts F and J should particularly be taken into account.

c) **Hot water systems.** When substantially replacing hot water systems, pipes and vessels – providing controls and insulation as if they are new installations. In lesser work, make reasonable provision for insulation, timing and thermostatic controls.

d) When replacing a complete lighting system serving more than 100m² of floor area, provide a new lighting system as if for a new building. Where only the complete luminaires are being replaced, provide new luminaires that meet the the standards given in paragraphs 1.43 or 1.48 (but the requirement does not apply where only components such as lamps or louvres are being replaced). Where only the control system is to be replaced, provide new controls that meet the standards in paragraphs 1.56 to 1.58 (but the requirement does not apply where only components such as switches and relays are being replaced).

e) **Air conditioning or mechanical ventilation systems.** When replacing air conditioning or mechanical ventilation systems that serve more than 200m² of floor area in office buildings, improving the Carbon Performance Rating in line with the guidance in paragraphs 1.62 to 1.66 of this Approved Document. In buildings other than offices, provide mechanical ventilation systems that meet the SFP standards in paragraph 1.67.

4.4 When carrying out work as described in paragraph 4.3 sub-clauses (b) to (e):

a) the work should be inspected and commissioned following the guidance in paragraph 2.6.

b) the building log-book should be prepared or updated as necessary to provide the appropriate details of the replacement controlled service or fitting (paragraphs 3.1 and 3.2).

c) the relevant part of the metering strategy should be prepared or revised as necessary, and additional metering provided where needed so as to enable the energy consumption of the replacement controlled service or fitting to be effectively monitored (paragraphs 3.3 to 3.6).

Material Alterations

4.5 "Material alterations" are defined in Regulation 3(2) as follows.

"An alteration is material for the purposes of these Regulations if the work, or any part of it, would at any stage result -

(a) in a building or controlled service or fitting not complying with a relevant requirement where previously it did; or

(b) in a building or controlled service or fitting which before the work commenced did not comply with a relevant requirement, being more unsatisfactory in relation to such a requirement."

4.6 "Relevant requirement" is defined in Regulation 3(3) as follows.

"In paragraph (2) "relevant requirement" means any of the following applicable requirements of Schedule 1, namely –

Part A (structure)

paragraph B1 (means of warning and escape)

paragraph B3 (internal fire spread - structure)

paragraph B4 (external fire spread)

paragraph B5 (access and facilities for the fire service)

Part M (access and facilities for disabled people)."

4.7 Reasonable provision where undertaking material alterations depends on the circumstances in the particular case and would need to take account of historic value (see paragraph 4.10 et seq). Without prejudice to the need for compliance with all the requirements in Schedule 1, the need to comply with the requirements of Parts F and J should particularly be taken into account. Possible ways of satisfying the requirements include:

a) **Roof insulation.** When substantially replacing any of the major elements of a roof structure - providing insulation to achieve the U-value for new buildings.

b) **Floor insulation.** Where the structure of a ground floor is to be substantially replaced – or re-boarded, providing insulation in heated rooms to the standard for new buildings.

c) **Wall insulation.** When substantially replacing complete exposed walls or their external renderings or cladding or internal surface finishes, or the internal surfaces of separating walls to unheated spaces, providing a reasonable thickness of insulation.

d) **Sealing measures.** When carrying out any of the above work, including reasonable sealing measures to improve airtightness.

e) **Controlled services and fittings.** When replacing controlled services and fittings, following the guidance in paragraph 4.3 and 4.4.

Material changes of use

4.8 "Material changes of use" are defined in Regulation 5 as follows.

... for the purposes of these Regulations, there is a material change of use where there is a change in the purposes for which or the circumstances in which a building is used, so that after that change -

(a) the building is used as a dwelling, where previously it was not;

(b) the building contains a flat, where previously it did not;

(c) the building is used as an hotel or a boarding house, where previously it was not;

(d) the building is used as an institution, where previously it was not;

(e) the building is used as a public building, where previously it was not;

(f) the building is not a building described in Classes I to VI in Schedule 2, where previously it was; or

(g) the building, which contains at least one dwelling, contains a greater or lesser number of dwellings than it did previously.

4.9 Reasonable provision where undertaking a material change of use depends on the circumstances in the particular case and would need to take account of historic value (see paragraph 4.10 et seq). Without prejudice to the need for compliance with all the requirements in Schedule 1, the need to comply with the requirements of Parts F and J should particularly be taken into account. Possible ways of satisfying the requirements include:

a) **Accessible lofts.** When upgrading insulation in accessible lofts, providing additional insulation to achieve a U-value not exceeding 0.25 W/m^2K where the existing insulation provides a U-value worse than 0.35 W/m^2K.

b) **Roof insulation.** When substantially replacing any of the major elements of a roof structure - providing insulation to achieve the U-value considered reasonable for new buildings.

c) **Floor insulation.** Where the structure of a ground floor is to be substantially replaced – providing insulation in heated rooms to the standard considered reasonable for new buildings.

d) **Wall insulation.** When substantially replacing complete exposed walls or their external renderings or cladding or internal surface finishes, or the internal surfaces of separating walls to unheated spaces, providing a reasonable thickness of insulation.

e) **Sealing measures.** When carrying out any of the above work, including reasonable sealing measures to improve airtightness.

f) **Controlled services and fittings.** When replacing controlled services and fittings, following the guidance in paragraphs 4.3 and 4.4.

Historic buildings

4.10 Historic buildings include -

a) listed buildings,

b) buildings situated in conservation areas,

c) buildings which are of architectural and historical interest and which are referred to as a material consideration in a local authority's development plan,

d) buildings of architectural and historic interest within national parks, areas of outstanding natural beauty, and world heritage sites.

4.11 The need to conserve the special characteristics of such historic buildings needs to be recognised: see BS 7913[55]. In such work the aim should be to improve energy efficiency where and to the extent that it is practically possible, always provided that the work does not prejudice the character of the historic building, or increase the risk of long-term deterioration to the building fabric or fittings. In arriving at an appropriate balance between historic building conservation and energy conservation, it would be appropriate to take into account the advice of the local planning authority's conservation officer.

4.12 Particular issues relating to work in historic buildings that warrant sympathetic treatment and where advice from others could therefore be beneficial include –

a) restoring the historic character of a building that had been subject to previous inappropriate alteration, e.g. replacement windows, doors and rooflights.

b) rebuilding a former historic building (e.g. following a fire or filling in a gap site in a terrace).

c) making provisions enabling the fabric to "breathe" to control moisture and potential long term decay problems: see SPAB Information Sheet No 4[56].

---

[55] BS 7913 The principles of the conservation of historic buildings, BSI, 1998 provides guidance on the principles that should be applied when proposing work on historic buildings.

[56] The need for old buildings to breathe, SPAB Information sheet 4, 1986.

Appendix A: Tables of U-values

Contents

Tables

Note: The values in these tables have been derived using the combined method, taking into account the effects of thermal bridging where appropriate. Intermediate values can be obtained from the tables by linear interpolation. As an alternative to using these tables, the procedures in Appendices B and C can be used to obtain a more accurate calculation of the thickness of insulation required.

Example calculations

Note: the examples are offered as indicating ways of meeting the requirements of Part L but designers also have to ensure that their designs comply with all the other parts of Schedule 1 to the Building Regulations.

Windows, doors and rooflights

The following tables provide indicative U-values for windows, doors and rooflights. Table A1 applies to windows and rooflights with wood or PVC-U frames. Table A2 applies to windows with metal frames, to which (if applicable) the adjustments for thermal breaks and rooflights in Table A3 should be applied. The tables do not apply to curtain walling or to other structural glazing not fitted in a frame. For the purposes of this Approved Document a roof window may be considered as a rooflight.

The U-value of a window or rooflight containing low-E glazing is influenced by the emissivity, ε_n, of the low-E coating. Low-E coatings are of two principal types, known as 'hard' and 'soft'. Hard coatings generally have emissivities in the range 0.15 to 0.2, and the data for ε_n = 0.2 should be used for hard coatings, or if the glazing is stated to be low-E but the type of coating is not specified. Soft coatings generally have emissivities in the range 0.05 to 0.1. The data for ε_n = 0.1 should be used for a soft coating if the emissivity is not specified.

When available, manufacturers' certified U-values (by measurement or calculation according to the standards given in Section 0) should be used in preference to the data given in these tables.

Table A1 Indicative U-values (W/m²·K) for windows and rooflights with wood or PVC-U frames, and doors

	Gap between panes			Adjustment for rooflights in dwellings[3]
	6mm	12mm	16mm or more	
Single glazing		4.8		+0.3
Double glazing (air filled)	3.1	2.8	2.7	+0.2
Double glazing (low-E, ε_n = 0.2)[1]	2.7	2.3	2.1	
Double glazing (low-E, ε_n = 0.15)	2.7	2.2	2.0	
Double glazing (low-E, ε_n = 0.1)	2.6	2.1	1.9	
Double glazing (low-E, ε_n = 0.05)	2.6	2.0	1.8	
Double glazing (argon filled)[2]	2.9	2.7	2.6	
Double glazing (low-E ε_n = 0.2, argon filled)	2.5	2.1	2.0	
Double glazing (low-E ε_n = 0.1, argon filled)	2.3	1.9	1.8	
Double glazing (low-E ε_n = 0.05, argon filled)	2.3	1.8	1.7	
Triple glazing	2.4	2.1	2.0	
Triple glazing (low-E, ε_n = 0.2)	2.1	1.7	1.6	
Triple glazing (low-E, ε_n = 0.1)	2.0	1.6	1.5	
Triple glazing (low-E, ε_n = 0.05)	1.9	1.5	1.4	
Triple glazing (argon filled)	2.2	2.0	1.9	
Triple glazing (low-E ε_n = 0.2, argon filled)	1.9	1.6	1.5	
Triple glazing (low-E ε_n = 0.1, argon filled)	1.8	1.4	1.3	
Triple glazing (low-E ε_n = 0.05, argon filled)	1.7	1.4	1.3	
Solid wooden door[4]		3.0		

Notes
[1] The emissivities quoted are normal emissivities. (Corrected emissivity is used in the calculation of glazing U-values.) Uncoated glass is assumed to have a normal emissivity of 0.89.
[2] The gas mixture is assumed to consist of 90% argon and 10% air.
[3] No correction need be applied to rooflights in buildings other than dwellings.
[4] For doors which are half-glazed the U-value of the door is the average of the appropriate window U-value and that of the non-glazed part of the door (e.g. 3.0W/m²K for a wooden door).

Table A2 Indicative U-values (W/m²·K) for windows with metal frames (4mm thermal break)

	gap between panes		
	6mm	12mm	16mm or more
Single glazing		5.7	
Double glazing (air filled)	3.7	3.4	3.3
Double glazing (low-E, ε_n = 0.2)	3.3	2.8	2.6
Double glazing (low-E, ε_n = 0.1)	3.2	2.6	2.5
Double glazing (low-E, ε_n = 0.05)	3.1	2.5	2.3
Double glazing (argon filled)	3.5	3.3	3.2
Double glazing (low-E, ε_n = 0.2,argon filled)	3.1	2.6	2.5
Double glazing (low-E, ε_n = 0.1, argon filled)	2.9	2.4	2.3
Double glazing (low-E, ε_n = 0.05, argon filled)	2.8	2.3	2.1
Triple glazing	2.9	2.6	2.5
Triple glazing (low-E, ε_n = 0.2)	2.6	2.2	2.0
Triple glazing (low-E, ε_n = 0.1)	2.5	2.0	1.9
Triple glazing (low-E, ε_n = 0.05)	2.4	1.9	1.8
Triple glazing (argon-filled)	2.8	2.5	2.4
Triple glazing (low-E, ε_n = 0.2, argon filled)	2.4	2.0	1.9
Triple glazing (low-E, ε_n = 0.1, argon filled)	2.2	1.9	1.8
Triple glazing (low-E, ε_n = 0.05, argon filled)	2.2	1.8	1.7

Note
For windows and rooflights with metal frames incorporating a thermal break other than 4mm, the following adjustments should be made to the U-values given in Table A2.

Table A3 Adjustments to U-values in Table A2 for frames with thermal breaks

Thermal break (mm)	Adjustment to U-value (W/m²K)	
	Window, or rooflight in building other than a dwelling	Rooflight in dwellings
0 (no break)	+0.3	+0.7
4	+0.0	+0.3
8	-0.1	+0.2
12	-0.2	+0.1
16	-0.2	+0.1

Note
Where applicable adjustments for both thermal break and rooflight should be made. For intermediate thicknesses of thermal breaks, linear interpolation may be used.

Corrections to U-values of roofs, walls and floors

Annex D of BS EN ISO 6946 provides corrections to U-values to allow for the effects of:

- air gaps in insulation
- mechanical fasteners penetrating the insulation layer
- precipitation on inverted roofs

The corrected U-value (U_c) is obtained by adding a correction term ΔU:

$$U_c = U + \Delta U$$

Table A4 gives the values of ΔU for some typical constructions.

Table A4 Corrections to U-values

	ΔU (W/m²K)
Roofs	
Insulation fixed with nails or screws	0.02
Insulation between joists or rafters	0.01
Insulation between and over joists or rafters	0.00
Walls	
Timber frame where the insulation partly fills the space between the studs	0.04
Timber frame where the insulation fully fills the space between the studs	0.01
Internal insulation fixed with nails or screws which penetrate the insulation	0.02
External insulation with metal fixings that penetrate the insulation	0.02
Insulated cavity wall with cavity greater than 75mm and tied with steel vertical-twist ties	0.02
Insulated cavity wall with a cavity less than or equal to 75mm tied with ties other than steel vertical-twist ties	0.00
Floors	
Suspended timber floor with insulation between joists	0.04

If the total ΔU is less than 3% of U then the corrections need not be applied and ΔU can be taken to be zero. However, where corrections are to be applied, before using the following tables the following steps should be carried out:

1) subtract ΔU from the desired U-value.

2) use this adjusted U-value in the tables when calculating the required thickness of insulation.

This thickness of insulation then meets the original desired U-value, having allowed for the ΔU correction(s).

Roofs

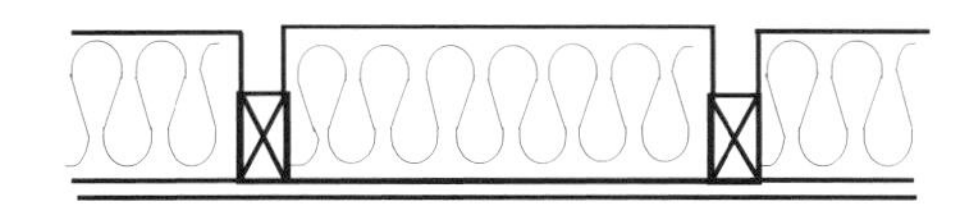

Table A5 Base thickness of insulation between ceiling joists or rafters

Design U-value (W/m²K)	Thermal conductivity of insulant (W/m·K) 0.020	0.025	0.030	0.035	0.040	0.045	0.050
	Base thickness of insulating material (mm)						
A	B	C	D	E	F	G	H
1 0.15	371	464	557	649	742	835	928
2 0.20	180	224	269	314	359	404	449
3 0.25	118	148	178	207	237	266	296
4 0.30	92	110	132	154	176	198	220
5 0.35	77	91	105	122	140	157	175
6 0.40	67	78	90	101	116	130	145

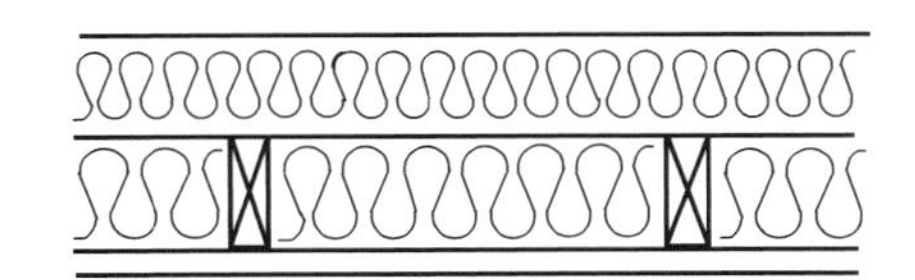

Table A6 Base thickness of insulation between and over joists or rafters

Design U-value (W/m²K)	Thermal conductivity of insulant (W/m·K) 0.020	0.025	0.030	0.035	0.040	0.045	0.050
	Base thickness of insulating material (mm)						
A	B	C	D	E	F	G	H
1 0.15	161	188	217	247	277	307	338
2 0.20	128	147	167	188	210	232	255
3 0.25	108	122	137	153	170	187	205
4 0.30	92	105	117	130	143	157	172
5 0.35	77	91	103	113	124	136	148
6 0.40	67	78	90	101	110	120	130

Note
Tables A5 and A6 are derived for roofs with the proportion of timber at 8%, corresponding to 48mm wide timbers at 600mm centres, excluding noggings. For other proportions of timber the U-value can be calculated using the procedure in Appendix B.

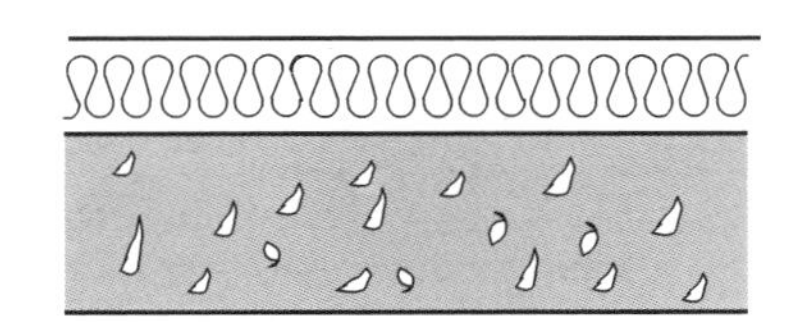

Table A7 Base thickness for continuous insulation

Design U-value (W/m²K)	Thermal conductivity of insulant (W/m·K) 0.020	0.025	0.030	0.035	0.040	0.045	0.050
	Base thickness of insulating material (mm)						
A	B	C	D	E	F	G	H
1 0.15	131	163	196	228	261	294	326
2 0.20	97	122	146	170	194	219	243
3 0.25	77	97	116	135	154	174	193
4 0.30	64	80	96	112	128	144	160
5 0.35	54	68	82	95	109	122	136
6 0.40	47	59	71	83	94	106	118

Table A8 Allowable reduction in base thickness for common roof components

	Concrete slab density (kg/m³)	Thermal conductivity of insulation (W/m·K) 0.020	0.025	0.030	0.035	0.040	0.045	0.050
		Reduction in base thickness of insulating for each 100mm of concrete slab						
	A	B	C	D	E	F	G	H
1	600	10	13	15	18	20	23	25
2	800	7	9	11	13	14	16	18
3	1100	5	6	8	9	10	11	13
4	1300	4	5	6	7	8	9	10
5	1700	2	2	3	3	4	4	5
6	2100	1	2	2	2	3	3	3
	Other materials and components	**Reduction in base thickness of insulating material (mm)**						
	A	B	C	D	E	F	G	H
7	10mm plasterboard	1	2	2	2	3	3	3
8	13mm plasterboard	2	2	2	3	3	4	4
9	13mm sarking board	2	2	3	3	4	4	5
10	12mm calcium silicate liner board	1	2	2	2	3	3	4
11	Roof space (pitched)	4	5	6	7	8	9	10
12	Roof space (flat)	3	4	5	6	6	7	8
13	19mm roof tiles	0	1	1	1	1	1	1
14	19mm asphalt (or 3 layers of felt)	1	1	1	1	2	2	2
15	50mm screed	2	3	4	4	5	5	6

Example 1: Pitched roof with insulation between ceiling joists or between rafters

Determine the thickness of the insulation layer required to achieve a U-value of 0.21W/m²K if insulation is between the joists, and 0.26 W/m²K if insulation is between the rafters. From Table A4 there is a ΔU correction of 0.01 W/m²K which applies to both the following cases. To allow for this, the 'look-up' U-value is reduced by 0.01W/m²K to 0.20 and 0.25W/m²K respectively.

For insulation placed between ceiling joists (look-up U-value 0.20W/m²K)

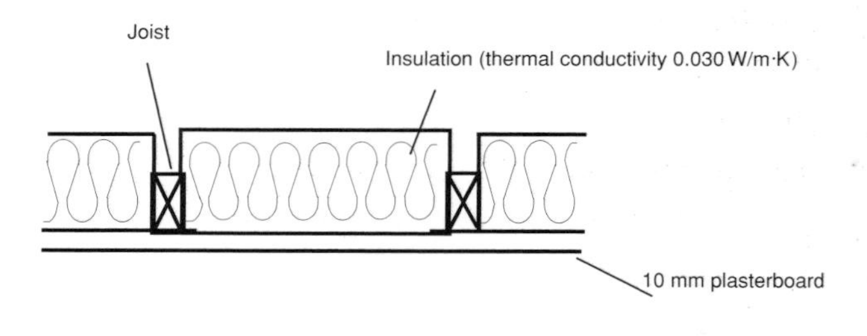

Using Table A5:

From **column D, row 2** of the table, the base thickness of insulation required is 269mm.

The base thickness may be reduced by taking account of the other materials as follows:

From Table A8:

19mm roof tiles	**column D, row 13**	= 1mm
Roofspace (pitched)	**column D, row 11**	= 6mm
10mm plasterboard	**column D, row 7**	= 2mm
Total reduction		= 9mm

The minimum thickness of the insulation layer between the ceiling joists required to achieve a U-value of 0.21W/m²K (including the ΔU correction) is therefore:

Base thickness less *total reduction*
i.e. 269 – 9 = **260mm**.

For insulation placed between rafters (look-up U-value 0.25W/m²K)

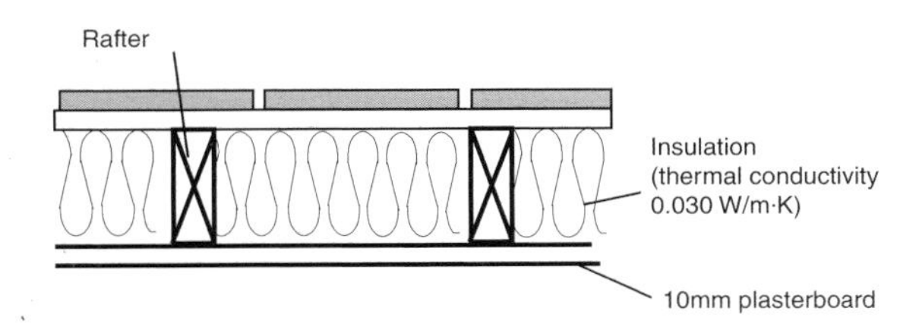

Using Table A5 :

From **column D, row 4** in the table, the base thickness of insulation required is 178mm.

The reductions in the base thickness are obtained as follows:

From Table A8:

19mm roof tiles	**column D, row 13**	= 1mm
10mm plasterboard	**column D, row 7**	= 2mm
Total reduction		= 3mm

The minimum thickness of the insulation layer between the rafters required to achieve a U-value of 0.25W/m²K (including the ΔU correction) is therefore:

Base thickness less *total reduction*
ie 178 - 3 = **175mm**.

Example 2: Pitched roof with insulation between and over ceiling joists

Determine the thickness of the insulation layer above the joists required to achieve a U-value of 0.20W/m²K for the roof construction shown below:

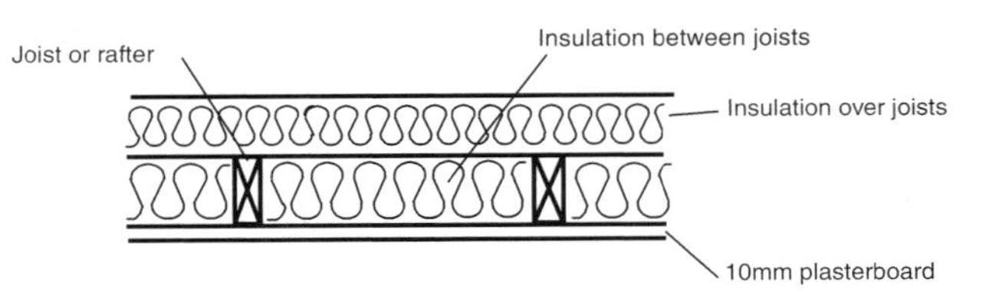

It is proposed to use mineral wool insulation between and over the joists with a thermal conductivity of 0.04W/mK.

Using Table A6:

From **column F, row 2** of the table, the base thickness of insulation layer = 210mm.

The base thickness may be reduced by taking account of the other materials as follows:

From Table A8:

19mm roof tiles	**column F, row 13**	= 1mm
Roofspace (pitched)	**column F, row 11**	= 8mm
10mm plasterboard	**column F, row 7**	= 3mm
Total reduction		= 12mm

The minimum thickness of the insulation layer over the joists, required in addition to the 100mm insulation between the joists, to achieve a U-value of 0.20W/m²K is therefore:

Base thickness less *total reduction*
ie 210 – 100 - 12 = **98mm**.

Example 3: Concrete deck roof

Determine the thickness of the insulation layer required to achieve a U-value of 0.25W/m²K for the roof construction shown below.

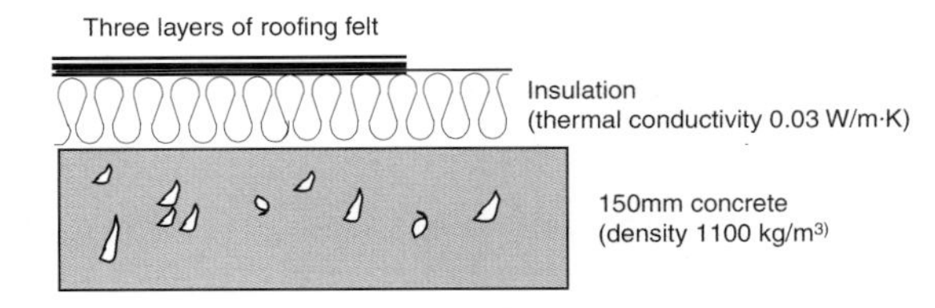

Using Table A7:

From **column D, row 3** of the table, the base thickness of the insulation layer is 116mm.

The base thickness may be reduced by taking account of the other materials as follows:

From Table A8:

3 layers of felt **column D, row 14** = 1mm

150mm concrete deck **column D, row 3** adjusted for 150mm thickness (1.5 x 8) = 12mm

Total reduction = 13mm

The minimum thickness of the insulation layer required to achieve a U-value of 0.25W/m²K is therefore:

Base thickness less *total reduction*
i.e. 116 - 13 = **103mm**.

Walls

Table A9 Base thickness of insulation layer

	Design U-value (W/m²K)	Thermal conductivity of insulant (W/m·K) 0.020	0.025	0.030	0.035	0.040	0.045	0.050
		Base thickness of insulating material (mm)						
	A	B	C	D	E	F	G	H
1	0.20	97	121	145	169	193	217	242
2	0.25	77	96	115	134	153	172	192
3	0.30	63	79	95	111	127	142	158
4	0.35	54	67	81	94	107	121	134
5	0.40	47	58	70	82	93	105	117
6	0.45	41	51	62	72	82	92	103

Table A10 Allowable reductions in base thickness for common components

	Component	Thermal conductivity of insulant (W/m·K) 0.020	0.025	0.030	0.035	0.040	0.045	0.050
		Reduction in base thickness of insulating material (mm)						
	A	B	C	D	E	F	G	H
1	Cavity (25mm or more)	4	5	5	6	7	8	9
2	Outer leaf brickwork	3	3	4	5	5	6	6
3	13mm plaster	1	1	1	1	1	1	1
4	13mm lightweight plaster	2	2	2	3	3	4	4
5	9.5mm plasterboard	1	2	2	2	3	3	3
6	12.5mm plasterboard	2	2	2	3	3	4	4
7	Airspace behind plasterboard drylining	2	3	4	4	5	5	6
8	9mm sheathing ply	1	2	2	2	3	3	3
9	20mm cement render	1	1	1	1	2	2	2
10	13mm tile hanging	0	0	0	1	1	1	1

Table A11 Allowable reductions in base thickness for concrete components

	Density (kg/m³)	Reduction in base thickness of insulation (mm) for each 100mm of concrete						
		Thermal conductivity of insulant (W/m·K)						
		0.020	0.025	0.030	0.035	0.040	0.045	0.050
	A	**B**	**C**	**D**	**E**	**F**	**G**	**H**
Concrete blockwork inner leaf								
1	600	9	11	13	15	17	20	22
2	800	7	9	10	12	14	16	17
3	1000	5	6	8	9	10	11	13
4	1200	4	5	6	7	8	9	10
5	1400	3	4	5	6	7	8	8
6	1600	3	3	4	5	6	6	7
7	1800	2	2	3	3	4	4	4
8	2000	2	2	2	3	3	3	4
9	2400	1	1	2	2	2	2	3
Concrete blockwork outer leaf or single leaf wall								
10	600	8	11	13	15	17	19	21
11	800	7	9	10	12	14	15	17
12	1000	5	6	7	8	10	11	12
13	1200	4	5	6	7	8	9	10
14	1400	3	4	5	6	6	7	8
15	1600	3	3	4	5	5	6	7
16	1800	2	2	3	3	3	4	4
17	2000	1	2	2	3	3	3	4
18	2400	1	1	2	2	2	2	3

Table A12 Allowable reductions in base thickness for insulated timber framed walls

	Thermal conductivity of insulation within frame (W/m·K)	Reduction in base thickness of insulation material (mm) for each 100mm of frame (mm)						
		Thermal conductivity of insulant (W/m·K)						
		0.020	0.025	0.030	0.035	0.040	0.045	0.050
	A	**B**	**C**	**D**	**E**	**F**	**G**	**H**
1	0.035	39	49	59	69	79	89	99
2	0.040	36	45	55	64	73	82	91

Note
The table is derived for walls for which the proportion of timber is 15%, which corresponds to 38mm wide studs at 600mm centres and includes horizontal noggings etc. and the effects of additional timbers at junctions and around openings. For other proportions of timber the U-value can be calculated using the procedure in Appendix B.

Example 4: Masonry cavity wall with internal insulation

Determine the thickness of the insulation layer required to achieve a U-value of 0.35W/m²K for the wall construction shown below.

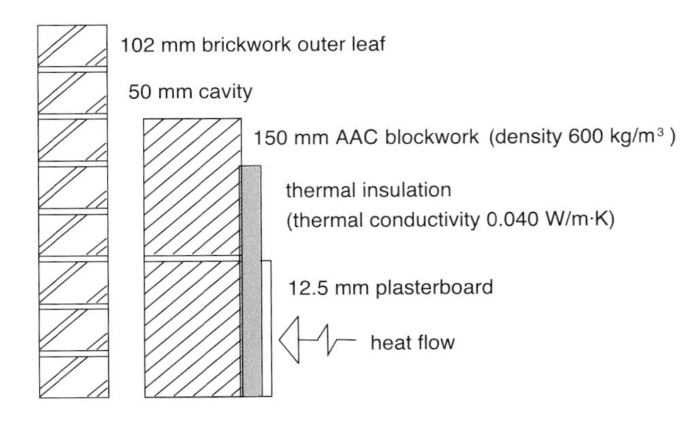

Using Table A9:

From **column F, row 4** of the table, the base thickness of the insulation layer is 107mm.

The base thickness may be reduced by taking account of the other materials as follows:

From Table A10:

Brickwork outer leaf	**column F, row 2**	= 5mm
Cavity	**column F, row 1**	= 7mm
Plasterboard	**column F, row 6**	= 3mm

And from table A11

Concrete blockwork	**column F, row 1** adjusted for 150mm block thickness (1.5 x 17)	= 26mm
Total reduction		= 41mm

The minimum thickness of the insulation layer required to achieve a U-value of 0.35W/m²K is therefore:

Base thickness less *total reduction*
i.e. 107 – 41 = **66mm.**

Example 5: Masonry cavity wall (tied with vertical-twist stainless-steel ties) filled with insulation with plasterboard on dabs

Determine the thickness of the insulation layer required to achieve a U-value of 0.37W/m²K for the wall construction shown below. From Table A4 there is a ΔU correction for the wall ties of 0.02W/m²K which applies. To allow for this, the 'look-up' U-value is reduced by 0.02W/m²K to 0.35W/m²K.

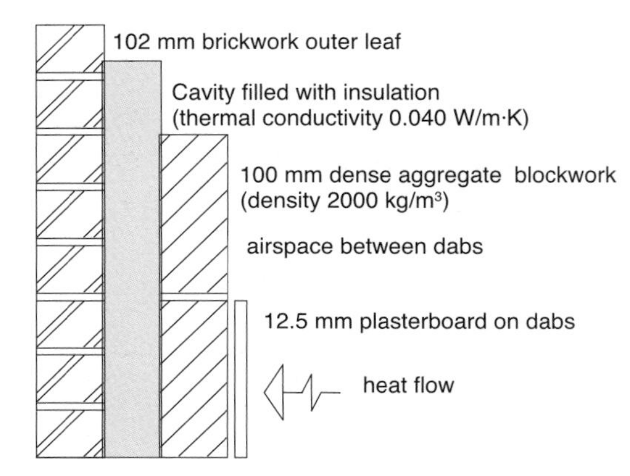

Using Table A9:

From **column F, row 4** of the table, the base thickness of the insulation layer is 107mm.

The base thickness may be reduced by taking account of the other materials as follows:

From Table A10:

Brickwork outer leaf	**column F, row 2**	= 5mm
Plasterboard	**column F, row 6**	= 3mm
Airspace behind plasterboard	**column F, row 7**	= 5mm

And from Table A11:

Concrete blockwork	**column F, row 1**	= 3mm
Total reduction		= 16mm

The minimum thickness of the insulation layer required to achieve a U-value of 0.37W/m²K (including ΔU for the wall ties) is therefore:

Base thickness less *total reduction* i.e. 107 – 16 = **91mm.**

Example 6: Masonry wall (tied with vertical-twist stainless-steel ties) with partial cavity-fill

Determine the thickness of the insulation layer required to achieve a U-value of 0.32W/m²K for the wall construction shown below. From Table A4 there is a ΔU correction for the wall ties of 0.02W/m²K which applies. To allow for this, the 'look-up' U-value is reduced by 0.02W/m²K to 0.30W/m²K.

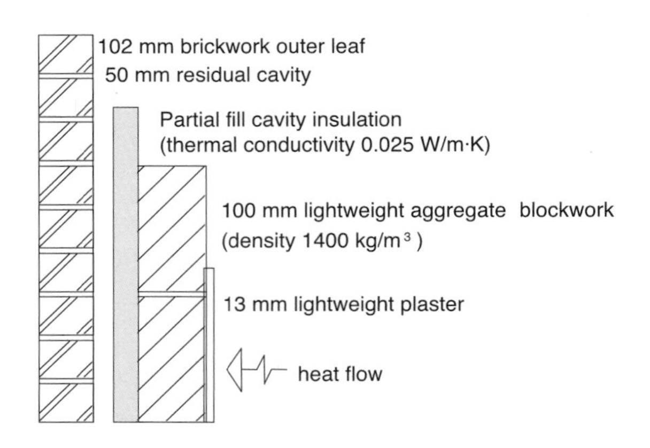

Using Table A9:

From **column C, row 3** of the table, the base thickness of the insulation layer is 79mm.

The base thickness may be reduced by taking account of the other materials as follows:

From Table A10:

Brickwork outer leaf	**column C, row 2**	= 3mm
Cavity	**column C, row 1**	= 5mm
Lightweight plaster	**column C, row 4**	= 2mm

And from Table A11:

Concrete blockwork	**column C, row 5**	= 4mm
Total reduction		= 14mm

The minimum thickness of the insulation layer required to achieve a U-value of 0.3W/m²K (including ΔU for the wall ties) is therefore:

Base thickness less *total reduction* i.e. 79 – 14 = **65mm.**

Example 7: Timber-framed wall

Determine the thickness of the insulation layer required to achieve a U-value of 0.35W/m²K for the wall construction shown below.

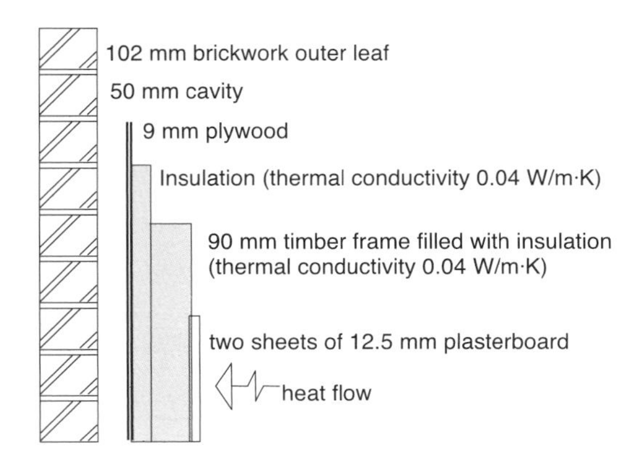

Using Table A9:

From **column F, row 4** of the table, the base thickness of the insulation layer is 107mm.

The base thickness may be reduced by taking account of the other materials as follows:

From Table A10:

Brickwork outer leaf	**column F, row 2**	= 5mm
Cavity	**column F, row 1**	= 7mm
Sheathing ply	**column F, row 8**	= 3mm
Plasterboard	**column F, row 6**	= 3mm
Plasterboard	**column F, row 6**	= 3mm

And from Table A12:

Timber frame	**column F, row 2** adjusted for stud thickness (73mm x 90/100)	= 66mm
Total reduction		= 87mm

The minimum thickness of the insulation layer required to achieve a U-value of 0.35W/m²K is therefore:

Base thickness less *total reduction*
i.e. 107 - 87 = **20mm.**

Ground floors

Note: in using the tables for floors it is first necessary to calculate the ratio P/A, where P is the floor perimeter length in metres and A is the floor area in square metres.

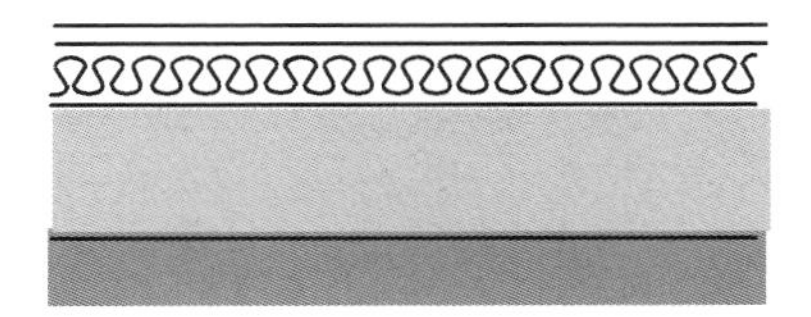

Table A13 **Insulation thickness for solid floors in contact with the ground**

		Insulation thickness (mm) for U-value of 0.20W/m²K						
		Thermal conductivity of insulant (W/m·K)						
	P/A	**0.020**	**0.025**	**0.030**	**0.035**	**0.040**	**0.045**	**0.050**
	A	**B**	**C**	**D**	**E**	**F**	**G**	**H**
1	1.00	81	101	121	142	162	182	202
2	0.90	80	100	120	140	160	180	200
3	0.80	78	98	118	137	157	177	196
4	0.70	77	96	115	134	153	173	192
5	0.60	74	93	112	130	149	167	186
6	0.50	71	89	107	125	143	160	178
7	0.40	67	84	100	117	134	150	167
8	0.30	60	74	89	104	119	134	149
9	0.20	46	57	69	80	92	103	115
		U-value of 0.25W/m²K						
10	1.00	61	76	91	107	122	137	152
11	0.90	60	75	90	105	120	135	150
12	0.80	58	73	88	102	117	132	146
13	0.70	57	71	85	99	113	128	142
14	0.60	54	68	82	95	109	122	136
15	0.50	51	64	77	90	103	115	128
16	0.40	47	59	70	82	94	105	117
17	0.30	40	49	59	69	79	89	99
18	0.20	26	32	39	45	52	58	65
		U-value of 0.30W/m²K						
19	1.00	48	60	71	83	95	107	119
20	0.90	47	58	70	81	93	105	116
21	0.80	45	56	68	79	90	102	113
22	0.70	43	54	65	76	87	98	108
23	0.60	41	51	62	72	82	92	103
24	0.50	38	47	57	66	76	85	95
25	0.40	33	42	50	59	67	75	84
26	0.30	26	33	39	46	53	59	66
27	0.20	13	16	19	22	25	28	32

Note
P/A is the ratio of floor perimeter (m) to floor area (m²).

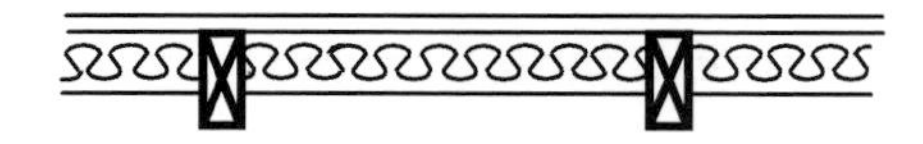

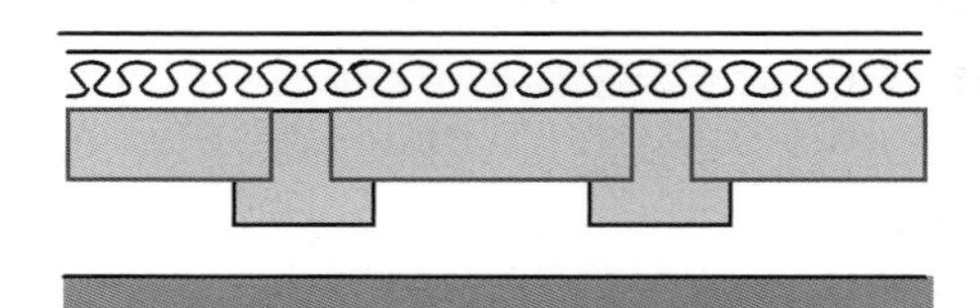

Table A14 Insulation thickness for suspended timber ground floors

	P/A	0.020	0.025	0.030	0.035	0.040	0.045	0.050
		Insulation thickness (mm) for U-value of 0.20W/m²K						
	A	B	C	D	E	F	G	H
1	1.00	127	145	164	182	200	218	236
2	0.90	125	144	162	180	198	216	234
3	0.80	123	142	160	178	195	213	230
4	0.70	121	139	157	175	192	209	226
5	0.60	118	136	153	171	188	204	221
6	0.50	114	131	148	165	181	198	214
7	0.40	109	125	141	157	173	188	204
8	0.30	99	115	129	144	159	173	187
9	0.20	82	95	107	120	132	144	156
		U-value of 0.25W/m²K						
10	1.00	93	107	121	135	149	162	176
11	0.90	92	106	119	133	146	160	173
12	0.80	90	104	117	131	144	157	170
13	0.70	88	101	114	127	140	153	166
14	0.60	85	98	111	123	136	148	161
15	0.50	81	93	106	118	130	142	154
16	0.40	75	87	99	110	121	132	143
17	0.30	66	77	87	97	107	117	127
18	0.20	49	57	65	73	81	88	96
		U-value of 0.30W/m²K						
19	1.00	71	82	93	104	114	125	135
20	0.90	70	80	91	102	112	122	133
21	0.80	68	78	89	99	109	119	129
22	0.70	66	76	86	96	106	116	126
23	0.60	63	73	82	92	102	111	120
24	0.50	59	68	78	87	96	104	113
25	0.40	53	62	70	79	87	95	103
26	0.30	45	52	59	66	73	80	87
27	0.20	28	33	38	42	47	51	56

Column headings B–H: Thermal conductivity of insulant (W/m·K).

Notes
P/A is the ratio of floor perimeter (m) to floor area (m²). The table is derived for suspended timber floors for which the proportion of timber is 12%, which corresponds to 48mm wide timbers at 400mm centres.

Table A15 Insulation thickness for suspended concrete beam and block ground floors

	P/A	0.020	0.025	0.030	0.035	0.040	0.045	0.050
		Insulation thickness (mm) for U-value of 0.20W/m²K						
	A	B	C	D	E	F	G	H
1	1.00	82	103	123	144	164	185	205
2	0.90	81	101	122	142	162	183	203
3	0.80	80	100	120	140	160	180	200
4	0.70	79	99	118	138	158	177	197
5	0.60	77	96	116	135	154	173	193
6	0.50	75	93	112	131	150	168	187
7	0.40	71	89	107	125	143	161	178
8	0.30	66	82	99	115	132	148	165
9	0.20	56	69	83	97	111	125	139
		U-value of 0.25W/m²K						
10	1.00	62	78	93	109	124	140	155
11	0.90	61	76	92	107	122	138	153
12	0.80	60	75	90	105	120	135	150
13	0.70	59	74	88	103	118	132	147
14	0.60	57	71	86	100	114	128	143
15	0.50	55	68	82	96	110	123	137
16	0.40	51	64	77	90	103	116	128
17	0.30	46	57	69	80	92	103	115
18	0.20	36	45	54	62	71	80	89
		U-value of 0.30W/m²K						
19	1.00	49	61	73	85	97	110	122
20	0.90	48	60	72	84	96	108	120
21	0.80	47	59	70	82	94	105	117
22	0.70	45	57	68	80	91	102	114
23	0.60	44	55	66	77	88	98	109
24	0.50	41	52	62	72	83	93	104
25	0.40	38	48	57	67	76	86	95
26	0.30	33	41	49	57	65	73	81
27	0.20	22	28	33	39	44	50	56

Column headings B–H: Thermal conductivity of insulant (W/m·K).

Note
P/A is the ratio of floor perimeter (m) to floor area (m²).

Example 8: Solid floor in contact with the ground

Determine the thickness of the insulation layer required to achieve a U-value of 0.3W/m²K for the ground floor slab shown below.

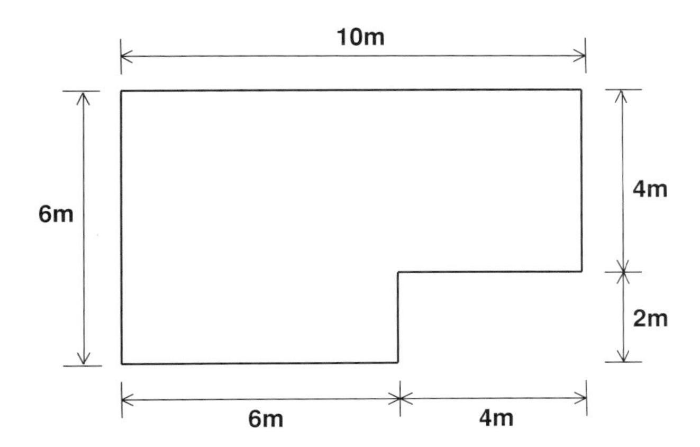

It is proposed to use insulation with a thermal conductivity of 0.025W/m·K.

The overall perimeter length of the slab is (10 + 4 + 4 + 2 + 6 + 6) = 32m.

The floor area of the slab is (6 x 6) + (4 x 4) = 52m².

The ratio:

$$\frac{\text{perimeter length}}{\text{floor area}} = \frac{32}{52} = 0.6$$

Using Table A13, **column C, row 23** indicates that **51mm** of insulation is required.

Example 9: Suspended timber floor

If the floor shown above was of suspended timber construction, the perimeter length and floor area would be the same, yielding the same ratio of:

$$\frac{\text{perimeter length}}{\text{floor area}} = \frac{32}{52} = 0.6$$

To achieve a U-value of 0.30W/m²·K, using insulation with a thermal conductivity of 0.04W/m·K, Table A14 **column F, row 23** indicates that the insulation thickness between the joists should be not less than **102mm**.

Upper floors

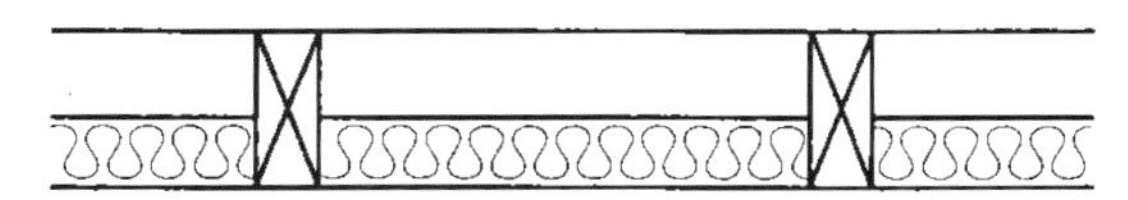

Table A16 Upper floors of timber construction

	Design U-value (W/m²K)	Thermal conductivity of insulant (W/m·K) 0.020	0.025	0.030	0.035	0.040	0.045	0.050
		Base thickness of insulation between joints to achieve design U-values						
	A	B	C	D	E	F	G	H
1	0.20	167	211	256	298	341	383	426
2	0.25	109	136	163	193	225	253	281
3	0.30	80	100	120	140	160	184	208

Note
Table A16 is derived for floors with the proportion of timber at 12% which corresponds to 48mm wide timbers at 400mm centres. For other proportions of timber the U-value can be calculated using the procedure in Appendix B.

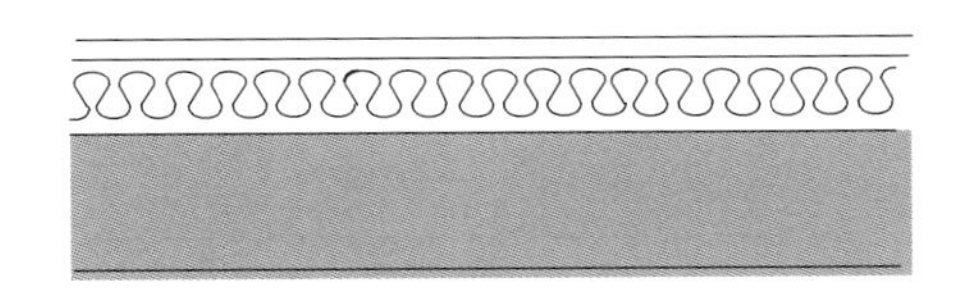

Table A17 Upper floors of concrete construction

	Design U-value (W/m²K)	Thermal conductivity of insulant (W/m·K) 0.020	0.025	0.030	0.035	0.040	0.045	0.050
		Base thickness of insulation to achieve design U-values						
	A	B	C	D	E	F	G	H
1	0.20	95	119	142	166	190	214	237
2	0.25	75	94	112	131	150	169	187
3	0.30	62	77	92	108	123	139	154

Table A18 Upper floors: allowable reductions in base thickness for common components

	Component	Thermal conductivity of insulant (W/m·K) 0.020	0.025	0.030	0.035	0.040	0.045	0.050
		Reduction in base thickness of insulating material (mm)						
	A	B	C	D	E	F	G	H
1	10mm plasterboard	1	2	2	2	3	3	3
2	19mm timber flooring	3	3	4	5	5	6	7
3	50mm screed	2	3	4	4	5	5	6

Building Materials

Table A19 Thermal conductivity of some common building materials

	Density (kg/m²)	Conductivity (W/m·K)
Walls		
Brickwork (outer leaf)	1700	0.77
Brickwork (inner leaf)	1700	0.56
Lightweight aggregate concrete block	1400	0.57
Autoclaved aerated concrete block	600	0.18
Concrete (medium density) (inner leaf)	1800	1.13
	2000	1.33
	2200	1.59
Concrete (high density)	2400	1.93
Reinforced concrete (1% steel)	2300	2.3
Reinforced concrete (2% steel)	2400	2.5
Mortar (protected)	1750	0.88
Mortar (exposed)	1750	0.94
Gypsum	600	0.18
	900	0.30
	1200	0.43
Gypsum plasterboard	900	0.25
Sandstone	2600	2.3
Limestone (soft)	1800	1.1
Limestone (hard)	2200	1.7
Fibreboard	400	0.1
Plasterboard	900	0.25
Tiles (ceramic)	2300	1.3
Timber (softwood), plywood, chipboard	500	0.13
Timber (hardwood)	700	0.18
Wall ties (stainless steel)	7900	17.0
Surface finishes		
External rendering	1300	0.57
Plaster (dense)	1300	0.57
Plaster (lightweight)	600	0.18
Roofs		
Aerated concrete slab	500	0.16
Asphalt	2100	0.70
Felt/bitumen layers	1100	0.23
Screed	1200	0.41
Stone chippings	2000	2.0
Tiles (clay)	2000	1.0
Tiles (concrete)	2100	1.5
Wood wool slab	500	0.10
Floors		
Cast concrete	2000	1.35
Metal tray (steel)	7800	50.0
Screed	1200	0.41
Timber (softwood), plywood, chipboard	500	0.13
Timber (hardwood)	700	0.18
Insulation		
Expanded polystyrene (EPS) board	15	0.040
Mineral wool quilt	12	0.042
Mineral wool batt	25	0.038
Phenolic foam board	30	0.025
Polyurethane board	30	0.025

Note
If available, certified test values should be used in preference to those in the table.

Appendix B: Calculating U-values

Introduction

B1 For building elements which contain repeating thermal bridges, such as timber joists between insulation in a roof or mortar joints around lightweight blockwork in a wall, the effect of thermal bridges should be taken into account when calculating the U-value. Other factors, such as metal wall ties and air gaps around insulation should also be included. The calculation method, known as the Combined Method, is set out in BS EN ISO 6946 and the following examples illustrate the use of the method for typical wall, roof and floor designs.

B2 In cases where the joists in roof, wall or floor constructions project beyond the surface of the insulation, the depths of the joists should be taken to be the same as the thickness of insulation for the purposes of the U-value calculation (as specified in BS EN ISO 6946).

B3 Thermal conductivity values for common building materials can be obtained from the CIBSE Guide Section A3 or from EN ISO 12524. For specific insulation products, however, data should be obtained from manufacturers.

B4 The procedure in this Appendix does not apply to elements containing metal connecting paths, for which the reader is directed to BRE IP 5/98 for metal cladding, CAB and CWCT guidance for curtain walls, and to BS EN ISO 10211-1 and -2 for other cases, and it does not deal with ground floors and basements (which are dealt with in Appendix C).

B5 The examples are offered as indicating ways of meeting the requirements of Part L but designers also have to ensure that their designs comply with all the other parts of Schedule 1 to the Building Regulations.

The procedure

B6 The U-value is calculated by applying the following steps:

a) Calculate the upper resistance limit (R_{upper}) by combining in parallel the total resistances of all possible heat-flow paths (i.e. sections) through the plane building element.

b) Calculate the lower resistance limit (R_{lower}) by combining in parallel the resistances of the heat flow paths of each layer separately and then summing the resistances of all layers of the plane building element.

c) Calculate the U-value of the element from $U = 1/R_T$,

$$\text{where } R_T = \frac{R_{upper} + R_{lower}}{2}$$

d) Adjust the U-value as appropriate to take account of metal fasteners and air gaps.

Example 1: Cavity wall with lightweight masonry leaf and insulated dry-lining

In this example there are two bridged layers - insulation bridged by timber and lightweight blockwork bridged by mortar (for a single bridged layer see the next example).

Diagram B1: Wall construction with two bridged layers

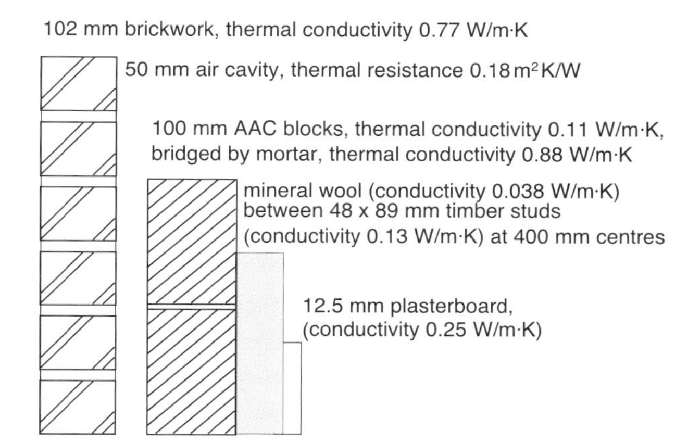

(Total thickness 353.5mm, $U = 0.32 W/m^2K$)

Layer	Material	Thickness (mm)	Thermal conductivity (W/m·K)	Thermal resistance (m²K/W)
	external surface	-	-	0.040
1	outer leaf brickwork	102	0.77	0.132
2	air cavity (unvented)	50	-	0.180
3(a)	AAC blocks (93%)	100	0.11	0.909
3(b)	mortar (7%)	(100)	0.88	0.114
4(a)	mineral wool (88%)	89	0.038	2.342
4(b)	timber battens (12%)	(89)	0.13	0.685
5	plasterboard	12.5	0.25	0.050
	internal surface	-	-	0.130

Upper resistance limit

There are four possible sections (or paths) through which heat can pass. The upper limit of resistance is therefore given by $R_{upper} = 1/(F_1/R_1 + ... + F_4/R_4)$ where F_m is the fractional area of section m and R_m is the total thermal resistance of section m. A conceptual illustration of the upper limit of resistance is shown in Diagram B2.

Diagram B2: Conceptual illustration of the upper limit of resistance

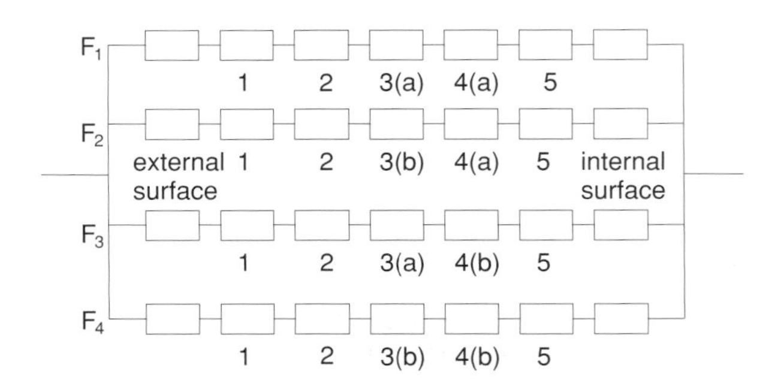

Resistance through section containing AAC blocks and mineral wool

External surface resistance	= 0.040
Resistance of brickwork	= 0.132
Resistance of air cavity	= 0.180
Resistance of AAC blocks	= 0.909
Resistance of mineral wool	= 2.342
Resistance of plasterboard	= 0.050
Internal surface resistance	= 0.130
Total thermal resistance R_1	= 3.783 m²K/W

Fractional area F_1= 93% x 88% = 0.818

Resistance through section containing mortar and mineral wool

External surface resistance	= 0.040
Resistance of brickwork	= 0.132
Resistance of air cavity	= 0.180
Resistance of mortar	= 0.114
Resistance of mineral wool	= 2.342
Resistance of plasterboard	= 0.050
Internal surface resistance	= 0.130
Total thermal resistance R_2	= 2.988 m²K/W

Fractional area F_2= 7% x 88% = 0.062

Resistance through section containing AAC blocks and timber

External surface resistance	= 0.040
Resistance of brickwork	= 0.132
Resistance of air cavity	= 0.180
Resistance of AAC blocks	= 0.909
Resistance of timber	= 0.685
Resistance of plasterboard	= 0.050
Internal surface resistance	= 0.130
Total thermal resistance R_3	= 2.126 m²K/W

Fractional area F_3= 93% x 12% = 0.112

Resistance through section containing mortar and timber

External surface resistance	= 0.040
Resistance of brickwork	= 0.132
Resistance of air cavity	= 0.180
Resistance of mortar	= 0.114
Resistance of timber	= 0.685
Resistance of plasterboard	= 0.050
Internal surface resistance	= 0.130
Total thermal resistance R_4	= 1.331 m²K/W

Fractional area F_4= 7% x 12% = 0.008

Combining these resistances we obtain:

$$R_{upper} = \frac{1}{\frac{F_1}{R_1}+\frac{F_2}{R_2}+\frac{F_3}{R_3}+\frac{F_4}{R_4}} = \frac{1}{\frac{0.818}{3.783}+\frac{0.062}{2.988}+\frac{0.112}{2.126}+\frac{0.008}{1.331}}$$

= 3.382 m²K/W.

Lower resistance limit

A conceptual illustration of the lower limit of resistance is shown in the Diagram B3.

Diagram B3: Conceptual illustration of the lower limit of resistance

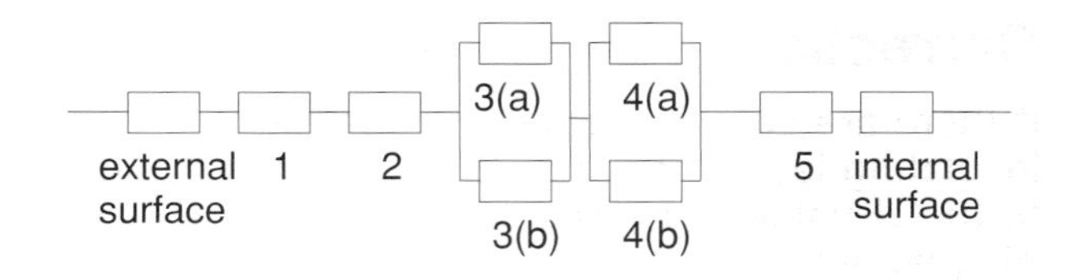

The resistances of the layers are added together to give the lower limit of resistance. The resistance of the bridged layer consisting of AAC blocks and mortar is calculated using:

$$R = \frac{1}{\frac{F_{blocks}}{R_{blocks}} + \frac{F_{mortar}}{R_{mortar}}}$$

and the resistance of the bridged layer consisting of insulation and timber is calculated using:

$$R = \frac{1}{\frac{F_{insul}}{R_{insul}} + \frac{F_{timber}}{R_{timber}}}$$

The lower limit of resistance is then obtained by adding together the resistances of the layers:

External surface resistance = 0.040

Resistance of brickwork = 0.132

Resistance of air cavity = 0.180

Resistance of first bridged layer (blocks and mortar)

$$= \frac{1}{\frac{0.93}{0.909} + \frac{0.707}{0.685}} = 0.611$$

Resistance of second bridged layer (insulation and timber)

$$= \frac{1}{\frac{0.88}{2.342} + \frac{0.12}{0.114}} = 1.815$$

Resistance of plasterboard = 0.050

Internal surface resistance = 0.130

Total (R_{lower}) = 2.958 m^2K/W

Total resistance of wall

The total resistance of the wall is the average of the upper and lower limits of resistance:

$$R_T = \frac{R_{upper} + R_{lower}}{2} = \frac{3.382 + 2.958}{2}$$

$= 3.170\ m^2K/W$.

Correction for air gaps

If there are small air gaps penetrating the insulating layer a correction should be applied to the U-value. The correction for air gaps is ΔU_g, where

$\Delta U_g = \Delta U'' \times (R_I / R_T)^2$

and where R_I is the thermal resistance of the layer containing gaps, R_T is the total resistance of the element and $\Delta U''$ is a factor which depends upon the way in which the insulation is fitted. In this example R_I is $1.815 m^2K/W$, R_T is $3.170 m^2K/W$ and $\Delta U''$ is 0.01 (ie correction level 1[1]). The value of ΔU_g is then

$\Delta U_g = 0.01 \times (1.815/3.170)^2 = 0.003 W/m^2K$.

U-value of the wall

The effect of air gaps or mechanical fixings[2] should be included in the U-value unless they lead to an adjustment in the U-value of less than 3%.

$U = 1/R_T + \Delta U_g$ (if ΔU_g is not less than 3% of $1/R_T$)

$U = 1/R_T$ (if ΔU_g is less than 3% of $1/R_T$)

In this case $\Delta Ug = 0.003\ W/m^2K$ and $1/R_T = 0.315\ W/m^2K$. Since ΔU_g is less than 3% of $(1/R_T)$,

$U = 1/R_T = 1/3.170 = 0.32 W/m^2K$ (expressed to two decimal places).

1 Applies for "Insulation installed in such a way that no air circulation is possible on the warm side of the insulation; air gaps may penetrate the insulation layer"

2 In this case the wall ties within the cavity do not penetrate any insulating layer and their effects need not be taken into account

Example 2: Timber framed wall

In this example there is a single bridged layer in the wall, involving insulation bridged by timber studs. The construction consists of outer leaf brickwork, a clear ventilated cavity, 10mm plywood, 38 x 140mm timber framing with 140mm of mineral wool quilt insulation between the timber studs and 2 sheets of plasterboard, each 12.5mm thick, incorporating a vapour check.

The timber fraction in this particular example is 15%. This corresponds to 38mm wide studs at 600mm centres and includes horizontal noggings etc. and the effects of additional timbers at junctions and around openings.

Diagram B4: Timber framed wall construction

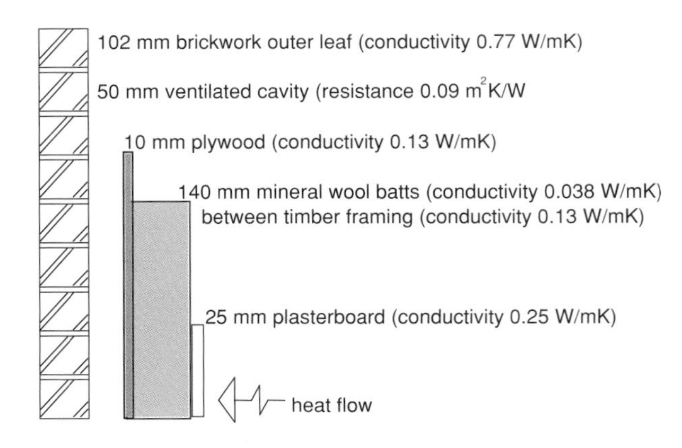

(Total thickness 327mm, U = 0.30W/m2K)

The thicknesses of each layer, together with the thermal conductivities of the materials in each layer, are shown below. The internal and external surface resistances are those appropriate for wall constructions. Layer 4 is thermally bridged and two thermal conductivities are given for this layer, one for the main part and one for the bridging part of the layer. For each homogeneous layer and for each section through a bridged layer, the thermal resistance is calculated by dividing the thickness (in metres) by the thermal conductivity.

Layer	Material	Thickness (mm)	Thermal conductivity (W/m·K)	Thermal resistance (m²K/W)
	external surface	-	-	0.040
1	outer leaf brick	102	0.77	0.132
2	ventilated air cavity	50	-	0.090
3	plywood	10	0.13	0.077
4(a)	mineral wool quilt between timber framing (85%)	140	0.038	3.684
4(b)	timber framing (15%)	(140)	0.13	1.077
5	plasterboard	25	0.25	0.100
	internal surface	-	-	0.130

Both the upper and the lower limits of thermal resistance are calculated by combining the alternative resistances of the bridged layer in proportion to their respective areas, as illustrated below. The method of combining differs in the two cases.

Upper resistance limit

When calculating the upper limit of thermal resistance, the building element is considered to consist of two thermal paths (or sections). The upper limit of resistance is calculated from:

$$R_{upper} = \frac{1}{\frac{F_1}{R_1} + \frac{F_2}{R_2}}$$

where F_1 and F_2 are the fractional areas of the two sections (paths) and R_1 and R_2 are the total resistances of the two sections. The method of calculating the upper resistance limit is illustrated conceptually in Diagram B5.

Diagram B5: Conceptual illustration of how to calculate the upper limit of thermal resistance

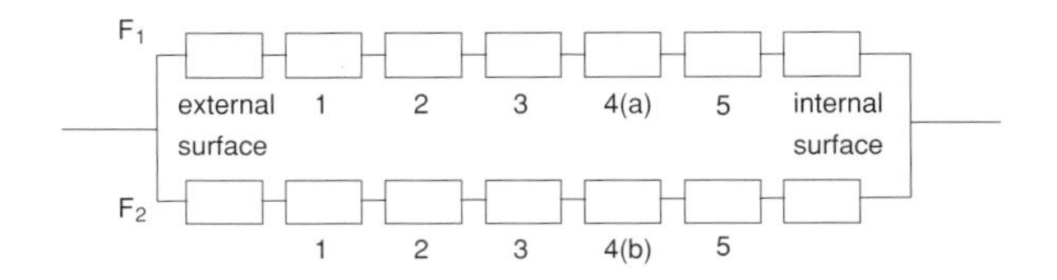

Resistance through the section containing insulation

External surface resistance	= 0.040
Resistance of bricks	= 0.132
Resistance of air cavity	= 0.090
Resistance of plywood	= 0.077
Resistance of mineral wool (85%)	= 3.684
Resistance of plasterboard	= 0.100
Internal surface resistance	= 0.130
Total (R_1)	= 4.253

Fractional area F_1 = 0.85 (85%)

The resistance through this section is therefore 4.253m²K/W.

Resistance through section containing timber stud

External surface resistance	= 0.040
Resistance of bricks	= 0.132
Resistance of air cavity	= 0.090
Resistance of plywood	= 0.077
Resistance of timber studs (15%)	= 1.077
Resistance of plasterboard	= 0.100
Internal surface resistance	= 0.130
Total (R_2)	= 1.646

Fractional area F_2 = 0.15 (15%)

The resistance through this section is therefore 1.646m²K/W

The upper limit of resistance is then:

$$R_{upper} = \frac{1}{\frac{F_1}{R_1} + \frac{F_2}{R_2}} = \frac{1}{\frac{0.85}{4.253} + \frac{0.15}{1.646}}$$

$$= 3.437\text{m}^2\text{K/W}.$$

Lower resistance limit

When calculating the lower limit of thermal resistance, the resistance of a bridged layer is determined by combining in parallel the resistances of the unbridged part and the bridged part of the layer. The resistances of all the layers in the element are then added together to give the lower limit of resistance.

The resistance of the bridged layer is calculated using:

$$R = \frac{1}{\frac{F_{insul}}{R_{insul}} + \frac{F_{timber}}{R_{timber}}}$$

The method of calculating the lower limit of resistance is illustrated conceptually in Diagram B6.

Diagram B6: Conceptual illustration of how to calculate the lower limit of thermal resistance

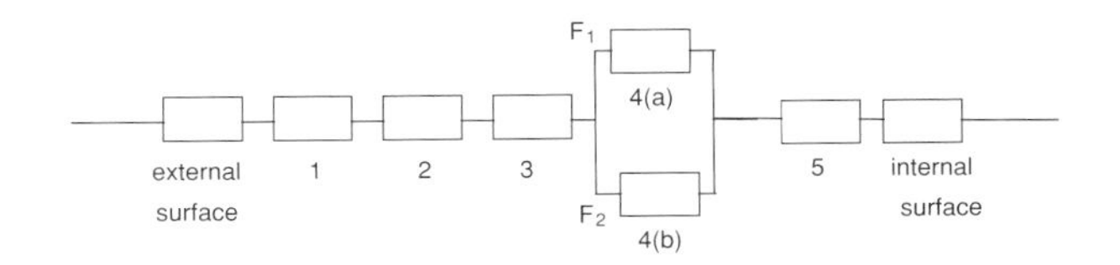

The lower limit of resistance is then obtained by adding up the resistances of all the layers:

External surface resistance	= 0.040
Resistance of bricks	= 0.132
Resistance of air cavity	= 0.090
Resistance of plywood	= 0.077

$$\text{Resistance of bridged layer} = \frac{1}{\frac{0.85}{3.684} + \frac{0.15}{1.077}} = 2.703$$

Resistance of plasterboard	= 0.100
Internal surface resistance	= 0.130
Total (R_{lower})	= 3.272

The lower limit of resistance is then 3.272m²K/W.

Total resistance of wall (not allowing for air gaps around the insulation)

The total resistance of the wall is the average of the upper and lower resistance limits:

$$R_T = \frac{R_{upper} + R_{lower}}{2} = \frac{3.437 + 3.272}{2}$$

$$= 3.354\text{m}^2\text{K/W}.$$

Correction for air gaps

If there are small air gaps penetrating the insulating layer a correction should be applied to the U-value to account for this. The correction for air gaps is ΔU_g, where

$\Delta U_g = \Delta U'' \times (R_I/R_T)^2$

and where R_I is the thermal resistance of the layer containing gaps, R_T is the total resistance of the element and $\Delta U''$ is a factor which depends upon the way in which the insulation is fitted. In this example R_I is $2.703m^2K/W$, R_T is $3.354m^2K/W$ and $\Delta U''$ is 0.01 (ie correction level 1). The value of ΔU_g is then

$\Delta U_g = 0.01 \times (2.703/3.354)^2 = 0.006W/m^2K$.

U-value of the wall

The effect of air gaps or mechanical fixings should be included in the U-value unless they lead to an adjustment in the U-value of less than 3%.

$U = 1/R_T + \Delta U_g$ (if ΔU_g is not less than 3% of $1/R_T$)

$U = 1/R_T$ (if ΔU_g is less than 3% of $1/R_T$)

In this case $\Delta U_g = 0.006W/m^2K$ and $1/R_T = 0.298W/m^2K$. Since ΔU_g is less than 3% of $(1/R_T)$,

$U = 1/R_T = 1 / 3.354 = 0.30W/m^2K$.

Appendix C: U-values of ground floors

C1 The guidance in this Approved Document states that a ground floor should not have a U-value exceeding 0.25 W/m²K if the Elemental Method of compliance is to be used. This can normally be achieved without the need for insulation if the perimeter to area ratio is less than 0.12 m/m² for solid ground floors or less than 0.09 m/m² for suspended floors. For most buildings, however, some ground floor insulation will be necessary to achieve this U-value or better performance. For exposed floors and for floors over unheated spaces the reader is referred to BS EN ISO 6946 or the CIBSE Guide Section A3.

C2 This Appendix provides a simple method for determining U-values which will suffice for most common constructions and ground conditions in the UK. More rigorous procedures are given in BS EN ISO 13370 and in CIBSE Guide Section A3 (1999 edition).

C3 For ground floors the U-value depends upon the type of soil beneath the building. Where the soil type is unknown, clay soil should be assumed as this is the most typical soil type in the UK. The tables which follow are based on this soil type. Where the soil is not clay or silt, the U-value should be calculated using the procedure in BS EN ISO 13370.

C4 Floor dimensions should be measured between finished internal faces of the external elements of the building including any projecting bays. In the case of semi-detached or terraced premises, blocks of flats and similar, the floor dimensions can either be taken as those of the premises themselves, or of the whole building. When considering extensions to existing buildings the floor dimensions may be taken as those of the complete building including the extension.

C5 Floor designs should prevent excessive thermal bridging at the floor edge so that the risk of condensation and mould are reasonably controlled. See BRE Report BR 262 *Thermal insulation: avoiding risks*.

C6 Unheated spaces outside the insulated fabric, such as attached garages or porches, should be excluded when determining the perimeter and area but the length of the wall between the heated building and the unheated space should be included when determining the perimeter.

C7 Data on U-values and insulation thicknesses for basements are given in the BCA/NHBC Approved Document "Basements for dwellings", ISBN 0-7210-1508-5.

Example of how to obtain U-values from the tables

The following example illustrates the use of the tables by interpolating between appropriate rows or columns.

A proposed building has a perimeter of 38.4 m and a ground floor area of 74.25 m². The floor construction consists of a 150mm concrete slab, 95mm of rigid insulation (thermal conductivity 0.04 W/m·K) and a 65mm screed. Only the insulation layer is included in the calculation of the thermal resistance.

Diagram C1

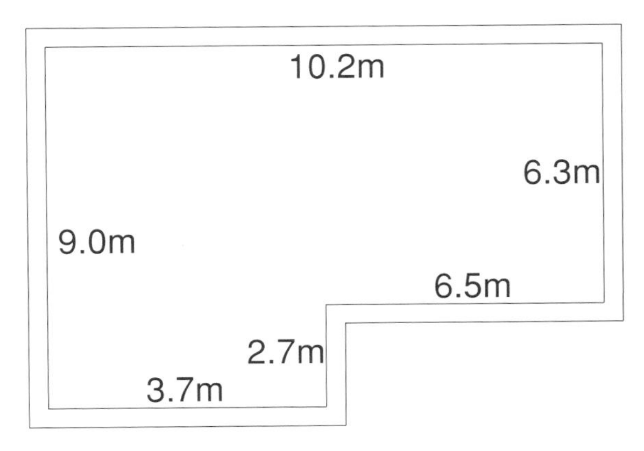

The perimeter to area ratio is equal to (38.4/74.25 = 0.517) m/m². Table C1 provides values for perimeter/area ratios of 0.50 and 0.55 but not for any values between 0.50 and 0.55. In this case, the U-value corresponding to a perimeter to area ratio of 0.50 should be used since 0.517 is closer to 0.50 than to 0.55.

The thermal resistance of the insulation is obtained by dividing the thickness (in metres) by the conductivity. The resistance is then 0.095/0.04 = 2.375 m²K/W.

The relevant part of table C1 is shown below:

Thermal resistance of all-over insulation (m²K/W)		
perimeter/area	**2.0**	**2.5**
0.50	0.28	0.24

The U-value corresponding to a thermal resistance of 2.375 m²K/W is obtained by linear interpolation as below:

$$U = 0.28 \times \frac{2.5 - 2.375}{2.5 - 2.0} + 0.24 \times \frac{2.375 - 2.0}{2.5 - 2.0}$$

$= 0.28 \times 0.25 + 0.24 \times 0.75$

$= 0.25 W/m^2K$

The U-value of this ground floor is therefore **0.25W/m²K**.

Note
In the example for Table C1 the appropriate row was chosen and interpolation was carried out between the appropriate columns. For all of the other tables, however, the appropriate column in the table should be selected and interpolation should be carried out between the appropriate rows.

Solid ground floors

Listed in Table C1 are U-values for solid ground floors. U-values are given in the following table for various perimeter-to-area ratios for a range of insulation levels. Where the floor is uninsulated the column corresponding to a thermal resistance of 0 should be used.

Table C1 U-values (W/m²K) for solid ground floors

	Thermal resistance of all-over insulation (m²K/W)					
perimeter/ area (m/m²)	**0**	**0.5**	**1**	**1.5**	**2**	**2.5**
0.05	0.13	0.11	0.10	0.09	0.08	0.08
0.10	0.22	0.18	0.16	0.14	0.13	0.12
0.15	0.30	0.24	0.21	0.18	0.17	0.15
0.20	0.37	0.29	0.25	0.22	0.19	0.18
0.25	0.44	0.34	0.28	0.24	0.22	0.19
0.30	0.49	0.38	0.31	0.27	0.23	0.21
0.35	0.55	0.41	0.34	0.29	0.25	0.22
0.40	0.60	0.44	0.36	0.30	0.26	0.23
0.45	0.65	0.47	0.38	0.32	0.27	0.23
0.50	0.70	0.50	0.40	0.33	0.28	0.24
0.55	0.74	0.52	0.41	0.34	0.28	0.25
0.60	0.78	0.55	0.43	0.35	0.29	0.25
0.65	0.82	0.57	0.44	0.35	0.30	0.26
0.70	0.86	0.59	0.45	0.36	0.30	0.26
0.75	0.89	0.61	0.46	0.37	0.31	0.27
0.80	0.93	0.62	0.47	0.37	0.32	0.27
0.85	0.96	0.64	0.47	0.38	0.32	0.28
0.90	0.99	0.65	0.48	0.39	0.32	0.28
0.95	1.02	0.66	0.49	0.39	0.33	0.28
1.00	1.05	0.68	0.50	0.40	0.33	0.28

Ground floors with edge insulation

Where horizontal or vertical edge insulation is used instead of all-over floor insulation, (P/A) x Ψ is added to the U-value to account for the effects of edge insulation, where P/A is the perimeter (m) to area (m²) ratio and Ψ is the edge insulation factor obtained from one of the following two tables. Since the term (P/A) x Ψ is negative it reduces the U-value of the ground floor. The tables apply only to floors without overall insulation.

Table C2 Edge insulation factor Ψ (W/m·K) for horizontal edge insulation

insulation width (m)	**thermal resistance of insulation (m²K/W)** 0.5	1.0	1.5	2.0
0.5	-0.13	-0.18	-0.21	-0.22
1.0	-0.20	-0.27	-0.32	-0.34
1.5	-0.23	-0.33	-0.39	-0.42

Table C3 Edge insulation factor Ψ (W/m·K) for vertical edge insulation

insulation depth (m)	**thermal resistance of insulation (m²K/W)** 0.5	1.0	1.5	2.0
0.25	-0.13	-0.18	-0.21	-0.22
0.50	-0.20	-0.27	-0.32	-0.34
0.75	-0.23	-0.33	-0.39	-0.42
1.00	-0.26	-0.37	-0.43	-0.48

For floors with both all-over insulation and edge insulation the calculation method in BS EN ISO 13370 can be used.

Uninsulated suspended ground floors

The following table gives U-values of uninsulated suspended floors for various perimeter to area ratios and for two levels of ventilation (expressed in m^2/m) below the floor deck. The data apply for the floor deck at a height not more than 0.5m above the external ground level where the wall surrounding the underfloor space is uninsulated.

Table C4 U-values (W/m^2K) of uninsulated suspended floors

perimeter to area ratio (m/m^2)	Ventilation opening area per unit perimeter of underfloor space	
	0.0015 m^2/m	0.0030 m^2/m
0.05	0.15	0.15
0.10	0.25	0.26
0.15	0.33	0.35
0.20	0.40	0.42
0.25	0.46	0.48
0.30	0.51	0.53
0.35	0.55	0.58
0.40	0.59	0.62
0.45	0.63	0.66
0.50	0.66	0.70
0.55	0.69	0.73
0.60	0.72	0.76
0.65	0.75	0.79
0.70	0.77	0.81
0.75	0.80	0.84
0.80	0.82	0.86
0.85	0.84	0.88
0.90	0.86	0.90
0.95	0.88	0.92
1.00	0.89	0.93

Insulated suspended floors

The U-value of an insulated suspended floor should be calculated using

$$U = 1/[(1/U_0) - 0.2 + R_f]$$

where U_0 is the U-value of an uninsulated suspended floor obtained using Table C4 or another approved method. R_f, the thermal resistance of the floor deck, is determined from U_f, the U-value of the floor deck, where

$$R_f = \frac{1}{U_f} - 0.17 - 0.17$$

and where U_f takes account of any thermal bridging in the floor deck and is calculated as recommended in BS EN ISO 6946 or by numerical modelling. The two values "0.17" are the two surface resistances.

Appendix D: Determining U-values for glazing

D1 Within the Elemental Method of compliance it is permissible to have windows, doors or rooflights with U-values that exceed the standard U-values provided that the average U-value of all of the windows (including rooflights) and doors taken together does not exceed the standard U-value in Table 1 in Section 1 of this Approved Document. The following example illustrates how this can be done.

D2 An office building is to have aluminium-framed windows of total area 682 m^2 and timber personnel doors of area 14 m^2. The proposed doors have a U-value of 3.3 W/m^2K which exceeds the standard U-value. The additional heat loss due to the higher U-value of the doors may be compensated for by lower window U-values.

D3 Using windows with a U-value of 1.9W/m^2K is sufficient to satisfy this requirement as shown in the following table and subsequent calculation.

Element	Area (m^2)	U-value (W/m^2K)	Rate of heat loss per degree (W/K)
Windows	682	2.1	1432.2
Doors	14	3.3	46.2
Total	696		1478.4

D4 The average U-value of the proposed windows and doors is 1478.4 ÷ 696, or 2.12 W/m^2K, which is below the standard U-value of 2.2W/m^2K when metal-framed windows are being used. The openings therefore satisfy the requirements of the Elemental Method.

Appendix E: Calculation examples

E1 This is an example of the procedure described in paragraphs from 1.14 to 1.16 (trade-off between construction elements), from 1.25 to 1.27 (heating system efficiency) and 1.32 (trade-off between construction elements and heating system efficiency).

E2 A detached, four storey office building 45m x 13m in plan and height 15m is to be constructed with glazing occupying 45% of the external wall area, using windows with a measured U-value of 2.0W/m²K. No rooflight glazing is proposed. The remaining exposed walls and the roof are to have U-values of 0.30W/m²K and 0.25W/m²K respectively, with the ground floor being insulated with 75mm expanded polystyrene with thermal resistance of all-over floor insulation 1.85m²K/W, giving a U-value of 0.20W/m²K (Appendix C). There is an unloading bay whose doors have area 27m² and the total area of personnel doors is 14m².

E3 The building is heated by a centralised heating plant with rated heat output of 90kW. The heating plant uses mains gas and consists of three regular boilers, each 80% efficient with a rated output of 30kW.

E4 The area of openings (windows and personnel doors) is greater than the 40% in Table 2. To compensate, the U-values of the walls and floor have been improved from the Elemental values given in Table 1: calculations are needed to demonstrate compliance. In addition, the heating system efficiency has to be checked for compliance with the requirements set in Table 5.

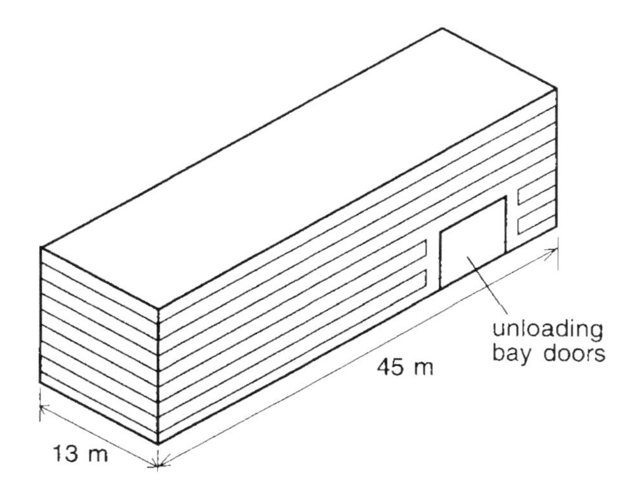

Proposed building

E5 Step 1 Calculate the areas of each element:

		Area (m²)
area of roof (45 x 13)	=	585
area of elevations		
(45 + 45 + 13 + 13) x 15	=	1740
area of windows (45% of 1740m²)	=	783
area of personnel doors	=	14
area of vehicle unloading bay doors	=	27
area of exposed wall		
(1740 - 783 - 14 - 27)	=	916
area of floor (45 x 13)	=	585

E6 Step 2 The rate of heat loss from the proposed building is calculated as follows:

Element	Area (m²)	U-value (W/m²K)	Rate of heat loss(W/K)
Roof	585	0.25	146.25
Exposed walls	916	0.3	274.80
Windows	783	2.0	1566.00
Personnel doors	14	2.0	28.00
Vehicle loading bay doors	27	0.7	18.90
Ground floor	585	0.2	117.00
	Total	**2910**	**2151.0**

Notional building

E7 For the notional building the area allowance is based on Table 2 and paragraph 1.15: that is 40% of the wall area plus 10% of the roof area (no more than half of the allowable rooflight area can be converted into an increased area of windows and doors).

E8 Step 1 Calculate the areas of each element:

		Area (m²)
area of roof lights (10% of 585m²)	=	58.5
area of roof (45 x 13) –58.5	=	526.5
area of elevations		
(45 + 45 + 13 + 13) x15	=	1740
area of windows and personnel		
doors (40% of 1740m²)	=	696
area of vehicle unloading bay doors	=	27
area of exposed wall (1740 - 696 - 27)	=	1017
area of floor (45 x 13)	=	585

E9 Step 2 Calculate the rate of heat loss from the notional building as follows:

Element	Area (m2)	U-value (W/m²K)	Rate of heat loss (W/K)
Roof lights	58.5	2.0	117.00
Roof	526.5	0.25	131.63
Exposed walls	1017	0.35	355.95
Windows and personnel doors	696	2.0	1392.00
Vehicle loading bay doors	27	0.7	18.90
Ground floor	585	0.25	146.25
Total	**2910**		**2161.7**

E10 The rate of heat loss from the proposed building is less than that from the notional building and therefore the requirements set out in Tables 1 and 2 in paragraphs 1.13 to 1.15 are satisfied.

E11 The efficiency of the heating system needs also to be considered. For the centralised heating plant a way of complying with the requirement would be to show that:

a) the carbon intensity of the heat generating equipment at the rated output of the heating system is not worse than the value shown in Table 5 column (a), and

b) the carbon intensity when the system is producing 30% of the rated output is not worse than the value shown in Table 5 column (b).

E12 The carbon intensity of the heat generating equipment at the rated output of the heating system

$$\varepsilon_c = \frac{1}{\sum R} \times \sum \frac{RC_f}{\eta_t}$$

where:

ε_c = the carbon intensity of the heating system (kg of carbon per kWh of useful heat);

R = the rated output of an individual element of heat raising plant (kW);

η_t = the gross thermal efficiency of that element of heat raising plant (kWh of heat per kWh of delivered fuel). For most practical cases, the efficiency may be taken as the full load efficiency for that element but where appropriate a part load efficiency based on manufacturer's certified data may be used as an alternative;

C_f = the carbon emission factor of the fuel supplying that element of heat raising plant (Table 6).

E13 The carbon intensity at 100% rated output (all boilers operating) is:

$$\varepsilon_c = \frac{1}{(30 + 30 + 30)} \times \left(\frac{30 \times 0.053}{0.80} + \frac{30 \times 0.053}{0.80} + \frac{30 \times 0.053}{0.80} \right)$$

$$= 0.0663\text{kg/kWh}$$

E14 At 100% rated output the calculated carbon intensity of 0.0663kg/kWh is not worse than the figure of 0.068kg/kWh given in Table 6 for natural gas, so the requirement is satisfied.

E15 The carbon intensity at 30% rated output, ie 0.3 x 90 = 27kW, when only the lead boiler is operating, is:

$$\varepsilon_c = \frac{1}{27} \times \frac{27 \times 0.053}{0.80} = 0.0663\text{kg/kWh}$$

E16 At 30% rated output the calculated carbon intensity of 0.0663kg/kWh is higher than the maximum of 0.065kg/kWh set in Table 5. This means that alternative solutions will need to be sought, for example to specify more efficient boiler(s) or improve envelope insulation standards to offset the higher carbon intensity of the heating system.

a) Making the lead boiler condensing.

E17 If the lead boiler is a condensing boiler with rated output of 30kW and 85% efficiency, then the carbon intensity at 100% rated output (three boilers operating) is:

$$\varepsilon_c = \frac{1}{(30 + 30 + 30)} \times \left(\frac{30 \times 0.053}{0.85} + \frac{30 \times 0.053}{0.80} + \frac{30 \times 0.053}{0.80} \right)$$

$$= 0.0650\text{kg/kWh}$$

(which is less than the maximum carbon intensity of 0.068kg/kWh).

E18 The carbon intensity at 30% rated output, when only the condensing boiler only is operating:

$$\varepsilon_c = \frac{1}{27} \times \frac{27 \times 0.053}{0.85} = 0.0624\text{kg/kWh}$$

E19 The calculated carbon intensity of the heating system is 0.0624kg/kWh, which is less than the maximum of 0.065kg/kWh set in Table 5. The requirement as to heating efficiency is now satisfied.

b) Improving the building envelope to offset the higher carbon intensity of the heating system

E20 As an alternative to the use of a condensing boiler, the average U-value of the envelope could be improved according to equation (3) in paragraph 1.32:

$$U_{req} = U_{ref} \frac{\varepsilon_{ref}}{\varepsilon_{act}}$$

where U_{ref} is obtained from the results of step 2 of the original calculation for the notional building:

$$U_{ref} = \frac{\text{Total rate of heat loss}}{\text{Total exposed surface area}} = \frac{2161.7}{2910.0}$$

$$= 0.743 \text{W/m}^2\text{K}$$

so that

$$U_{req} = 0.743 \times \frac{0.743}{0.0663} = 0.728 \text{W/m}^2\text{K}.$$

E21 One possibility of achieving this reduction in average U-value is a slight reduction in window area. If the window area is reduced from 783m² to 763m² (with a commensurate increase in exposed wall area), the heat loss calculation becomes.

Element	Area (m²)	U-value (W/m²K)	Rate of heat loss (W/K)
Roof	585	0.25	146.25
Exposed walls	936	0.3	280.80
Windows (~44%)	763	2.0	1526.00
Personnel doors	14	2.0	28.00
Vehicle loading bay doors	27	0.7	18.90
Ground floor	585	0.2	117.00
Total	**2910**		**2117.0**

E22 The average U-value is now 2117.0/2910 = 0.727W/m²K. This does not exceed U_{req} (0.728W/m²K), so although the carbon intensity of the heating system is higher than the figures in Table 5, that is sufficiently compensated by the fabric as permitted by paragraph 1.32.

Appendix F: Meeting the lighting standards

General lighting in office, industrial and storage buildings

By selection of lamp and luminaire types

F1 The performance standard for the electric lighting system in these building types depends on the efficiencies of both the lamp/ballast combination and the luminaire. The recommendation in paragraph 1.43 is met if:

a) the installed lighting capacity in circuit Watts comprises lighting fittings incorporating lamps of the type shown in Table F1, and

b) all the luminaires have a light output ratio of at least 0.6.

F2 A maximum of 500W of installed lighting in the building is exempt from the above requirement (paragraph 1.46).

Table F1 **Types of high efficacy lamps for non-daylit areas of offices, industrial and storage buildings**

Light source	Types
High pressure sodium	All ratings above 70W
Metal halide	All ratings above 70W
Tubular fluorescent	All 26mm diameter (T8) lamps and 16mm diameter (T5) lamps rated above 11W, provided with low-loss or high frequency control gear.
Compact fluorescent	All ratings above 26W

F3 Otherwise, if the use of other types of lighting or less efficient luminaires is planned, a calculation of the average initial luminaire efficacy is required (paragraph 1.44).

Example calculation of average luminaire efficacy

F4 A small industrial unit is being constructed incorporating production, storage and office areas. Lighting in the production area (which is non-daylit) is to be controlled by staged time switching to coincide with shift patterns. The storage area is anticipated to be occasionally visited, and is to be controlled by local absence detection, where a sensor switches the lighting off if no one is present, but switching on is done manually. The office areas are daylit; the furthest luminaire is less than 6m from the window wall, which is 30% glazed with clear low emissivity double glazing. Lighting control in this area is by localised infra red switch. Lighting control in the non-daylit corridor, toilet and foyer areas is by full occupancy sensing with automatic on and off.

F5 The lighting controls therefore meet the requirements of paragraph 1.56 (for the office and storage areas) and paragraph 1.58 (for the production and circulation areas).

F6 Table F2 below shows a schedule of the light sources proposed, together with a calculation of the overall average luminaire efficacy. It incorporates the luminaire control factor, which allows for the reduced energy use due to lighting in daylit and rarely occupied spaces. The storage areas are occasionally visited and incorporate absence detection, so have a luminaire control factor of 0.8.

F7 The daylit office areas with local manual switching also have a luminaire control factor of 0.8. Note that if the office areas had tinted glazing, of transmittance 0.33, the equivalent area of glazing of transmittance 0.7 would need to be calculated. This is 30% x 0.33/0.7 = 14% of the window wall area. As this area is less than 20% of the window wall, the office areas would not count as daylit if this type of glazing were used.

Table F2

Position	Number N	Description	Circuit Watts (W) per fitting	Lamp lumen output Ø (lm) per fitting	Luminaire light output ratio LOR	Luminaire control factor C_L	Total corrected luminaire output =Nx Ø x LOR/C_L(lm)	Total circuit Watts (W)
Production	16	250W high bay metal halide	271	17000	0.8	1	217600	4336
Offices	12	4 x 18W fluorescent with aluminium Cat 2 louvre and high frequency control gear	73	4600	0.57	0.8	39330	876
Storage	16	58W fluorescent with aluminium louvres and mains frequency control gear	70	4600	0.6	0.8	55200	1120
Circulation, toilets and foyer	30	24W compact fluorescent mains frequency downlights	32	1800	0.4	1	21600	960
						Totals	**333730**	**7292**

F8 From Table F2, the total corrected lumen output of all the lamps in the installation is 333,730 lumens.

F9 The total circuit Watts of the installation is 7292 Watts. Therefore the average luminaire efficacy is 333,730/7292 = 45.8 lumens/Watt. As this is greater than 40 lumens/Watt, the proposed lighting scheme therefore meets the requirements of this Approved Document. Note that up to 500W of any form of lighting, including lamps in luminaires for which light output ratios are unavailable, could also be installed in the building according to paragraph 1.46.

General lighting in other building types

Lighting calculation procedure to show average circuit efficacy is not less than 50 lumens/watt

F10 A lighting scheme is proposed for a new public house comprising a mixture of concealed perimeter lighting using high frequency fluorescent fittings and supplementary tungsten lamps in the dining area. Lights in the dining and lounge areas are to be switched locally from behind the bar. Lighting to kitchens and toilets is to be switched locally.

F11 Table F3 shows a schedule of the light sources proposed together with the calculation of the overall average circuit efficacy.

Table F3

Position	Number	Description	Circuit Watts (W) per lamp	Lumen output (lm) per lamp	Total circuit Watts (W)	Total Lamp lumen output (lm)
Over tables	20	60W tungsten	60	710	1200	14,200
Concealed perimeter and bar lighting	24	32W T8 fluorescent high frequency ballast	36	3300	864	79,200
Toilets and circulation	6	18W compact fluorescent mains frequency ballast	23	1200	138	7,200
Kitchens	6	50W, T8 fluorescent high frequency ballast	56	5200	336	31,200
				Totals	2538	131,800

F12 From Table F3, the total lumen output of the lamps in the installation is 131,800 lumens.

F13 The total circuit Watts of the installation is 2538 Watts.

F14 Therefore, the average circuit efficacy is:

$$\frac{131800}{2538} = 51.9 \text{ lumens/Watt}$$

F15 The proposed lighting scheme therefore meets the requirements of this Approved Document.

F16 If 100W tungsten lamps were used in the dining area instead of the 60W lamps actually proposed, the average circuit efficacy would drop to 43.4 lumens/W, which is unsatisfactory. If, however, 11W compact fluorescent lamps, which have similar light output to 60W tungsten lamps, were used in the dining area the average circuit efficacy would be 83.2 lumens/W.

Lighting calculation procedure to show that 95% of installed circuit power is comprised of lamps listed in table 8 (paragraph 1.48)

F17 A new hall and changing rooms are to be added to an existing community centre. The proposed lighting scheme incorporates lamps that are listed in Table 8 except for some low voltage tungsten halogen downlighters which are to be installed in the entrance area with local controls. A check therefore has to be made to show that the low voltage tungsten halogen lamps comprise less than 5% of the overall installed capacity of the lighting installation.

Main hall

F18 Twenty wall mounted uplighters with 250W high pressure Sodium lamps are to provide general lighting needs. The uplighters are to be mounted 7m above the floor. On plan, the furthest light is 20.5m from its switch, which is less than three times the height of the light above the floor.

F19 It is also proposed to provide twenty 18W compact fluorescent lights as an additional system enabling instant background lighting whenever needed.

Changing rooms, corridors and entrance

F20 Ten 58W, high frequency fluorescent light fittings are to be provided in the changing rooms and controlled by occupancy detectors. Six more 58W fluorescent light fittings are to be located in the corridors and the entrance areas and switched locally. Additionally, in the entrance area there are to be the six 50W tungsten halogen downlighters noted above.

Calculation

F21 A schedule of light fittings is prepared as follows:

Position	Number	Description of light source	Circuit Watts per lamp	Total circuit Watts (W)
Main hall	20	250W SON	286W	5720
Main hall	20	18W compact fluorescent	23W	460
Entrance, changing rooms and corridors	16	58W HF fluorescent	64W	1024
Entrance	6	50W low voltage tungsten halogen	55W	330
			Total =	7534W

F22 The percentage of circuit Watts consumed by lamps not listed in Table 8 is

$$\frac{330 \times 100}{7534} = 4.4\ \%$$

F23 Therefore, more than 95% of the installed lighting capacity, in circuit Watts, is from light sources listed in Table 8. The switching arrangements comply with paragraph 1.58. The proposed lighting scheme therefore meets the requirements of the Regulations.

Appendix G: Methods for office buildings

Assessing the contribution to carbon emissions due to building services design and operation

G1 The efficiencies of buildings, and of the services systems that produce the indoor conditions required by occupants, can be assessed and compared provided a consistent system is used to describe the buildings and their energy use.

G2 Applying such a consistent approach in the office building sector has allowed energy consumption benchmarks to be developed with which the performance of existing buildings, or the likely performance of new designs, can be compared. The benchmarks result from a number of surveys of operational buildings, and are included in Energy Consumption Guide 19 "Energy use in offices" (ECON 19).

Performance benchmarks

G3 The information contained in ECON 19 provides benchmarks for the energy consumed by ACMV, heating and lighting services, together with benchmark information describing the hours of use of the equipment. Benchmarks also describe the energy consumed by the additional equipment necessary to support use of the building for typical office activities. The benchmarks refer to office buildings described as representing 'typical' and 'good practice' for the sector.

Design Assessment

G4 The annual energy likely to be consumed by a particular service can be estimated as the product of the total installed input power rating of the plant installed to provide the service and the annual hours of use of that plant at the equivalent of full load. The annual hours of use can be considered to be the result of combining a benchmark value for the 'typical' hours of use of the service with a management factor that acts to reduce or increase this value. The management factor is a number related to the provisions that have been included that have the potential to help the occupier control and manage the use of the plant.

G5 The Carbon Performance Rating (CPR) referred to in paragraphs G10 to G19 of this Appendix is a technique for assessing the likely performance of building services systems using this design information. It uses benchmarks consistent with ECON19 and is intended to estimate the potential for efficient operation of building services systems using information available at the design or construction stage.

Performance assessment

G6 The inclusion of meters (Section 3 of this Approved Document) improves the confidence with which occupiers may assess their buildings' performance by estimating the energy consumed by servicing plant and the additional equipment required for the full operation of the building.

G7 A technique of estimating operational energy consumption, and comparing the achieved performance of buildings with the ECON19 benchmarks, has been developed to assess the achieved performance of office buildings. This method is described in CIBSE Technical Memorandum TM 22 "Energy Assessment and Reporting Methodology: Office Assessment Method".

G8 A means of comparing the design of services with benchmarks of installed load and energy use is described the CIBSE Guide volume "Energy Efficiency in Buildings".

G9 The results of ongoing performance assessment could be used to provide valuable information from which to maintain and improve performance benchmarks, and hence the CPR method, and to inform the design process.

The Carbon Performance Rating (CPR)

The CPR for mechanical ventilation - $CPR_{(MV)}$

G10 The assessment is based on the calculation of a Carbon Performance Rating using the following relationship:

$$CPR_{(MV)} = PD \times HD \times CD \times FD$$

G11 The design is considered to represent acceptable practice where the result of the calculation $CPR_{(MV)}$ = 6.5 or less.

G12 For the system installed to provide mechanical ventilation, the factors PD, HD, CD and FD are as defined below:

PD is the total installed capacity (sum of the input kW ratings) of the fans installed to provide mechanical ventilation divided by the relevant treated area (square metres)

HD is the typical annual equivalent hours of full load operation, and is taken as 3700 hours per year

CD is the conversion factor relating the emissions of carbon to the fuel used, here electricity, in kgC/kWh. (See Table 6 for carbon emission factors)

FD is a factor which depends on the provisions that are made to control and manage the installed plant and which could act to improve the annual efficiency of the plant above that of the typical installation, or to reduce the effective annual hours of use. (See Table G1)

The CPR for air conditioning – $CPR_{(MR)}$

G13 The assessment is based on the calculation of a Carbon Performance Rating using the following relationship:

$$CPR_{(ACMV)} = (PD \times HD \times CD \times FD) + (PR \times HR \times CR \times FR)$$

G14 The design is considered to represent acceptable practice where the result of the calculation $CPR_{(ACMV)}$ = 10.3 or less.

G15 For the distribution system transferring cooled medium to the conditioned spaces, the factors PD, HD, CD and FD are as defined below:

PD is the total installed capacity (sum of the input kW ratings) of the fans and pumps installed to distribute air and/or cooled media around the building divided by the relevant treated area (square metres)

HD is the typical annual equivalent hours of full load operation, and is taken as 3700 hours per year

CD is the conversion factor relating the emissions of carbon to the fuel used, here electricity, in kgC/kWh. (See Table 6 for carbon emission factors)

FD is a factor which depends on the provisions that are made to control and manage the installed plant and which could act to improve the annual efficiency of the plant above that of the typical installation, or to reduce the effective annual hours of use. (See Table G2)

For the refrigeration system, the factors PR, HR, CR and FR are as defined below:

PR is the total installed capacity (sum of the input kW ratings) of the plant installed to provide the cooling or refrigeration function divided by the relevant treated area (square metres)

HR is the typical annual equivalent hours of full load operation of the refrigeration plant, and is taken as 1000 hours per year

CR is the conversion factor relating the emissions of Carbon to the fuel used, here most frequently electricity, sometimes gas, in kgC/kWh. (See Table 6 for carbon emission factors)

FR is a factor which depends on the provisions that are made to control and manage the installed plant and which could act to improve the annual efficiency of the plant above that of the typical installation, or to reduce the effective annual hours of use. (See Table G3)

Plant control and management factors

G16 Tables G1, G2 and G3 below itemise a number of control and management features which could act to improve the annual efficiency of the relevant plant above that of the typical installation, or to reduce the effective annual hours of use. Values to be associated with each feature are obtained from column A, B or C as appropriate and the resultant factor is obtained by multiplying together all of the individual values obtained. Values are selected from columns A, B and C of the table depending on the extent to which facilities for monitoring and reporting are provided, as follows:

Column C — No monitoring provided

Column B — Provision of energy metering of plant and/or metering of plant hours run, and/or monitoring of internal temperatures in zones

Column A — Provision as B above, plus the ability to draw attention to 'out of range' values.

G17 The plant management features for Table G1 are:

Table G1 To obtain factor (FD) for the air distribution system

Plant management features	Values A	B	C
a) Operation in mixed mode with natural ventilation	0.85	0.9	0.95
b) Controls which restrict the hours of operation of distribution system	0.9	0.93	0.95
c) Efficient means of controlling air flow rate	0.75	0.85	0.95
Column product (FD):			

a) Mixed mode operation available as a result of including sufficient openable windows to provide the required internal environment from natural ventilation when outdoor conditions permit. This may only apply where the perimeter zone exceeds 80% of the treated floor area.

b) Control capable of limiting plant operation to occupancy hours with the exceptions noted below in which operation outside the hours of occupancy forms a necessary part of the efficient use of the system:

for control of condensation,

for optimum start/stop control, or

as part of a 'night cooling' strategy.

c) Air flow rate controlled by a variable motor speed control which efficiently reduces input power at reduced output; variable pitch fan blades. (Damper, throttle or inlet guide vane controls do not attract this factor).

Table G2 To obtain factor (FD) for the cooling distribution system

Plant management features	Values A	B	C
a) Operation in mixed mode with natural ventilation	0.85	0.9	0.95
b) Controls which restrict the hours of operation of distribution system	0.9	0.93	0.95
c) Efficient means of controlling air flow rate	0.75	0.85	0.95
Column product (FD):			

G18 The plant management features for Table G2 are:

a) Mixed mode operation available as a result of including sufficient openable windows to provide the required internal environment from natural ventilation when outdoor conditions permit. This may only apply where the perimeter zone exceeds 80% of the treated floor area. This factor is credited only where interlocks are provided to inhibit the air conditioning supply in zones with opened windows.

b) and c) are as described in Table G1 above for mechanical ventilation.

Table G3 To obtain factor (FR) for the refrigeration plant

Plant management features	Values A	B	C
a) Free cooling from cooling tower	0.9	0.93	0.95
b) Variation of fresh air using economy cycle or mixed mode operation	0.85	0.9	0.95
c) Controls to restrict hours of operation	0.85	0.9	0.95
d) Controls to prevent simultaneous heating and cooling in the same zone	0.9	0.93	0.95
e) Efficient control of plant capacity, including modular plant	0.9	0.93	0.95
f) Partial ice thermal storage	1.8	1.86	1.9
g) Full ice thermal storage	0.9	0.93	0.95
Column product (FR):			

G19 The plant management features for Table G3 are:

a) Systems that permit cooling to be obtained without the operation of the refrigeration equipment when conditions allow (eg 'strainer cycle'; 'thermosyphon').

b) Systems that incorporate an economy cycle in which the fresh air and recirculated air mix is controlled by dampers, or where mixed mode operation is available as defined below Table G2.

c) Controls that are capable of limiting plant operation to the hours of occupancy of the building, with the exceptions noted below in which operation outside the hours of occupancy forms a necessary part of the efficient use of the system:

for control of condensation,

for optimum start/stop control, or

as part of a strategy to pre-cool the building overnight using outside air.

d) Controls that include an interlock or dead band capable of precluding simultaneous heating and cooling in the same zone.

e) Refrigeration plant capacity controlled on-line by means that reduce input power in proportion to cooling demand and maintain good part load efficiencies (eg. modular plant with sequence controls; variable speed compressor). (Hot gas bypass control does not attract this factor).

f) Partial ice storage in which the chiller is intended to operate continuously, charging the store overnight and supplementing its output during occupancy.

g) Full ice storage in which the chiller operates only to recharge the thermal store overnight and outside occupancy hours.

Example CPR calculations

Example calculation for a office proposal including air conditioning

G20 In this example it is intended to include an air conditioning system in a new office building. The relevant details from the proposal are that:

The total area to be treated by the system is $3000m^2$.

Cooling will be provided by two speed-controlled electrically powered compressors, with a total rated input power of 150kW.

The refrigeration compressor energy consumption will be metered.

The fans used to distribute cooled air to treated spaces have a total rated input power of 35kW.

The fan energy consumption will be metered.

A time clock control is to be provided so that the operation of the cooling system (refrigeration and air distribution) may be restricted to occupancy hours.

Windows in treated areas will be openable so that natural ventilation may be used, and the cooling system turned off, when required.

The CPR calculation for air conditioning is:

$CPR_{(ACMV)}$ = (PD x HD x CD x FD) + (PR x HR x CR x FR)

In this proposal, for the cooling distribution system:

PD is the total installed capacity (sum of the input kW ratings) of the fans divided by the relevant treated area (square metres)

= 0.0117 (35/3000)

HD = 3700 hours per year

CD is the carbon conversion factor for electricity, in kgC/kWh. (See Table 6 for carbon emission factors)

= 0.113

FD = 0.84, determined from Table G2 as follows:

As the major plant will be metered, factors from Column B of the Table are used. Then:

Factor for including the opportunity for natural ventilation (mixed mode operation) = 0.9

Factor for including provision to restrict the hours of use of the system (time control) = 0.93

Column product (FD) = 0.84 (0.9 x 0.93)

And, for the refrigeration system:

PR = the total installed capacity (sum of the input kW ratings) of the refrigeration plant divided by the treated area (square metres)

= 0.05 (150/3000)

HR = 1000 hours per year

CR = the carbon conversion factor for electricity, in kgC/kWh. (See Table 6 for carbon emission factors)

= 0.113

FR = 0.75, determined from Table G3 as follows:

As the major plant will be metered, factors from Column B of the Table are used. Then:

Factor for including the opportunity for natural ventilation (mixed mode operation) = 0.9

Factor for including provision to restrict the hours of use of the system (time control) = 0.9

Factor for providing efficient means of controlling plant capacity = 0.93

Column product (FR) = 0.75 (0.9 x 0.9 x 0.93)

The CPR calculation is then:

$CPR_{(ACMV)}$ = (0.0117 x 3700 x 0.113 x 0.84) + (0.05 x 1000 x 0.113 x 0.75)

= 8.35

The proposal therefore achieves a calculated rating of 8.35, which is lower than the required CPR of 10.3 and would therefore be acceptable on this basis.

Note: The rating of 8.35 indicates that, under similar patterns of occupancy and use, the proposed building would be likely to cause about 20% less carbon emission than would be caused by one like the typical air conditioned office building defined in ECON 19.

Example calculation for a proposal to increase the area treated by an office mechanical ventilation system

G21 In this example it is intended to increase the area treated by an existing office mechanical ventilation system. The relevant details from the proposal are that:

The total area to be treated by the system is to be increased from $3200m^2$ to $3800m^2$.

The total input power rating of the fans is to be unchanged at 72kW.

The fan input power will be metered, where previously it was not.

An existing time clock control provision for the system is to be kept.

The CPR calculation for mechanical ventilation is:

$CPR_{(MV)}$ = (PD x HD x CD x FD)

In this proposal, for the existing air distribution system:

PD is the total installed capacity (sum of the input kW ratings) of the fans divided by the relevant treated area (square metres)

= 0.0225 (72/3200)

HD = 3700 hours per year

CD is the carbon conversion factor for electricity, in kgC/kWh. (See Table 6 for carbon emission factors)

= 0.113

FD = 0.95, determined from Table G1 as follows:

As the plant is not metered, factors from Column C of the Table are used. Then:

Factor for including provision to restrict the hours of use of the system (time control) = 0.95

Column product (FD) = 0.95

The CPR calculation is then:

$CPR_{(MV)}$ = (0.0225 x 3700 x 0.113 x 0.95)
= 8.94

Since this calculated rating of 8.94 for the existing system is higher than the target 6.5 for new construction, the altered system would be required to reduce the rating by 10%, or to 7.15 (6.5+10%), whichever is the less demanding. In this case the 10% reduction is the less demanding and results in a target rating of 8.04.

In this proposal, for the extended air distribution system:

PD is the total installed capacity (sum of the input kW ratings) of the fans divided by the relevant treated area (square metres)

= 0.01895 (72/3800)

HD = 3700 hours per year

CD is the carbon conversion factor for electricity, in kgC/kWh. (See Table 6 for carbon emission factors)

= 0.113

FD = 0.93, determined from Table G1 as follows:

As plant will now be metered, factors from Column B of the Table are used, then:

Factor for including provision to restrict the hours of use of the system (time control) = 0.93

Column product (FD) = 0.93

The CPR calculation is then:

$CPR_{(MV)}$ = (0.01895 x 3700 x 0.113 x 0.93)
= 7.37

This alteration achieves a rating lower than its particular target of 8.04 and would therefore be acceptable on this basis.

Note: The rating of 7.37 indicates that, under similar patterns of occupancy and use, the proposed building would be likely to cause about 13% greater carbon emission than one like the typical air conditioned office building defined in ECON 19.

Example calculation for a office proposal including air conditioning and a dedicated, air conditioned, computer suite.

G22 In this example it is intended to include an air conditioning system in a new office building that also houses a dedicated computer suite that will be served as a separate controlled zone from the centralised air conditioning system. The relevant details from the proposal are that:

The total area to be treated by the system is $3500m^2$.

Cooling will be provided by two speed-controlled electrically powered compressors, with a total rated input power of 225kW.

The refrigeration compressor energy consumption will be metered.

The fans used to distribute cooled air to treated spaces have a total rated input power of 45kW.

The fan energy consumption will be metered.

Windows in treated office areas will be openable, with interlocks to disable the local air conditioning terminals, so that natural ventilation may be used when required.

Efficient fan speed control will be installed to accommodate the variations in demand for air supply.

The treated area of the computer room is $500m^2$. The designer has estimated that the refrigeration input power required to service the computer room is 45kW and the fan input power required to provide air supply to the computer room is 10kW.

The relevant plant input power ratings and the relevant treated area are the result of subtracting the computer room area and the plant required to service it from the totals, hence:

Relevant treated area = $3000m^2$
(3500 – 500)

Relevant refrigeration installed capacity
= 180kW (225 – 45)

Relevant fan installed capacity
= 35kW (45 – 10)

The CPR calculation for air conditioning is:

$CPR_{(ACMV)}$ = (PD x HD x CD x FD)
+ (PR x HR x CR x FR)

In this proposal, for the cooling distribution system:

PD is the total installed capacity (sum of the input kW ratings) of the fans divided by the relevant treated area (square metres)

= 0.0117 (35/3000)

HD = 3700 hours per year

CD is the carbon conversion factor for electricity, in kgC/kWh. (See Table 6 for carbon emission factors)

= 0.113

FD = 0.765, determined from Table G2 as follows:

As the major plant will be metered, factors from Column B of the Table are used. Then:

Factor for including the opportunity for natural ventilation (mixed mode operation) = 0.9

Factor for including efficient control of air flow rate = 0.85

Column product (FD) = 0.765 (0.9 x 0.85)

And, for the refrigeration system:

PR = the total installed capacity (sum of the input kW ratings) of the refrigeration plant divided by the treated area (square metres)

= 0.06 (180/3000)

HR = 1000 hours per year

CR = the carbon conversion factor for electricity, in kgC/kWh. (See Table 6 for carbon emission factors)

= 0.113

FR = 0.837, determined from Table G3 as follows:

As the major plant will be metered, factors from Column B of the Table are used. Then:

Factor for including the opportunity for natural ventilation (mixed mode operation) = 0.9

Factor for providing efficient means of controlling plant capacity = 0.93

Column product (FR) = 0.837 (0.9 x 0.93)

The CPR calculation is then:

$CPR_{(ACMV)}$ = (0.0117 x 3700 x 0.113 x 0.765) + (0.06 x 1000 x 0.113 x 0.837) = 9.42

The proposal therefore achieves a calculated rating of 9.42, which is lower than the required CPR of 10.3 and would therefore be acceptable on this basis.

Note: The rating of 9.42 indicates that, under similar patterns of occupancy and use, the proposed building would be likely to cause about 10% less carbon emission than one like the typical air conditioned office building defined in ECON 19.

Appendix H: Methods for solar overheating

H1 This appendix provides the detail for the procedure described in paragraph 1.23a).

H2 When estimating the solar load, the space being considered should be split into perimeter and interior zones. Perimeter zones are those defined by a boundary drawn a maximum of 6m away from the window wall(s). Interior zones are defined by the space between this perimeter boundary and the non-window walls or the perimeter boundary of another perimeter zone. When calculating the average solar cooling load, the contribution from all windows within that zone should be included, plus the area of any rooflight (or part rooflight) that is within the zone boundary. For interior zones, the contribution from all rooflights (or part rooflight) that is within its zone boundary should be included. For each zone within the space, the total solar cooling load (Qslw+ Qslr) should be no greater than 25W/m². The average solar cooling load per unit floor area averaged between the hours of 07:30 and 17:30 can be calculated by use of the following equations.

a) The contribution from vertical glazing should be calculated from

$$Q_{slw} = \frac{1}{A_p} \sum A_g q_s f_c (1-f_{rw}) \qquad \text{(H1)}$$

where:-

Q_{slw} is the solar load per unit floor area (W/m²).

A_P is the floor area of the perimeter zone (m²).

A_g is the area of the glazed opening (m²).

q_s is the solar load for the particular orientation of opening (W/m² of glazing) - Table H1.

f_c is a correction factor for glazing/blind combination (Table H2).

f_{rw} is the framing ratio for the window (default value for vertical windows = 0.1).

b) The contribution from any horizontal rooflights in the space should be calculated from

$$Q_{slr} = q_{sr}\, g_{rr}\, f_c (1-f_{rr}) \qquad \text{(H2)}$$

where:-

Q_{slr} is the solar load per unit floor area (W/m² of floor area).

q_{sr} is the solar load for horizontal openings (W/m² of opening area) (Table H1).

g_{rr} is the ratio of the total area of rooflight to the floor area.

f_{rr} is the framing ratio for the rooflight (default value for horizontal rooflights =0.3).

f_s is a correction factor for glazing/blind combination (Table H2).

f_c is a correction factor for glazing/blind combination (Table H2).

Table H1 Average solar load between 07.30 and 17.30 for different orientations

Orientation	Average solar load (W/m²)
N	125
NE/NW	160
E/W	205
SE/SW	198
S	156
Horizontal	327

Table H2 Correction factors for intermittant shading using various glass/blind combinations

Glazing/blind combination (described from inside to outside)	Correction factor f_C
Blind/clear/clear	0.95
Blind/clear/reflecting	0.62
Blind/clear/absorbing	0.66
Blind/low-e/clear	0.92
Blind/low-e/reflecting	0.60
Blind/low-e/absorbing	0.62
Clear/blind/clear	0.69
Clear/blind/reflecting	0.47
Clear/blind/absorbing	0.50
Clear/clear/blind/clear	0.56
Clear/clear/blind/reflecting	0.37
Clear/clear/blind/absorbing	0.39
Clear/clear/blind	0.57
Clear/clear/clear/blind	0.47

H3 As a preferred alternative to the generic numbers in these tables, shading coefficient data for a particular device can be used:

a) For fixed shading (including units with absorbing or reflecting glass), the correction factor fc is given by

$$f_c = \frac{S_c}{0.7} \qquad \text{(H3)}$$

b) For moveable shading, the correction factor is given by

$$f_c = \frac{1 + \frac{S_c}{0.7}}{2} \qquad \text{(H4)}$$

where S_c is the shading coefficient for the glazing/shading device combination, i.e. the ratio of the instantaneous heat gain at normal incidence by the glazing/shading combination relative to that transmitted by a sheet of 4mm clear glass.

c) Where there is a combination of fixed and moveable shading, the correction factor is given by

$$f_c = \frac{S_{cf} + S_{ctot}}{1.4} \qquad \text{(H5)}$$

where S_{cf} is the shading coefficient of the fixed shading (with glazing) and S_{ctot} is the shading coefficient of the combination of glazing and fixed and moveable shading.

Example 1

H4 Consider a classroom in a school. The room is 9m long by 6m deep, with a floor to ceiling height of 2.9m. There is glazing on one wall, with rooflights along the internal wall opposite the window wall. The windows are 1200mm wide by 1000mm high, and there are six such windows in the external wall, which faces SE. The windows are clear double glazed, with mid-pane blinds, of wooden frames with a framing percentage of 25%. There are three 0.9m^2 horizontal rooflights, with an internal blind and low-e glass on the inner pane of the double pane unit.

For the windows Q_{slw}=(6 x 1.2 x 1.0 x 198 x 0.69 x (1-0.25))/(9*6) = 13.7 W/m^2

For the rooflight Q_{slr} = (327 x (3 x 0.9/54) x 0.92 x 0.7) = 10.5 W/m^2

The total solar load is 13.7 + 10.5 = 24.2, which is less than the limiting value of 25 W/m^2.

Example 2

H5 This example shows how the method in this appendix could be used to determine, for each space, a shading coefficient that would enable the solar loads to meet the requirements of Part L.

H6 Consider an office building, with a floor to ceiling height of 2.8m and curtain walling construction with a glazing ratio of 0.65. The default framing factor of 0.1 is appropriate in this case. The long side of the office faces south and the short side faces west. The main office area is open plan, but there is a 5m by 3m corner office. For the open plan areas, the perimeter zone is defined by the 6m depth rule, but for the corner office, it is defined by the partitions. In order to avoid solar overheating, it is proposed to provide fixed external shading. Equations (H1) and (H3) can be used to determine the required shading coefficient for the glass/shading combination*. Combining equations (H1) and (H3) and re-arranging gives:-

$$S_c = \frac{0.7 \times Q_{slw} \times A_p}{\sum A_g \times q_s \times (1 - f_{rw})} \qquad \text{(H6)}$$

For the south facing open plan area, consider a typical module 10m wide. In this case

S_c = (0.7 x 25 x 10 x 6)/(10 x 2.8 x 0.65 x 156 x 0.9) = 0.41.

For a similar west facing open plan area

S_c = (0.7 x 25 x 10 x 6)/(10 x 2.8 x 0.65 x 205 x 0.9) = 0.31.

* Note that a different equation would be obtained if a moveable shading device were in use, since equation (H1) would then need to be combined with equation (H4).)

For the corner office, there are two window orientations, and so the summation term of equation (H5) is calculated as shown in the table.

Orientation of window	Area of window (m^2)	Solar load per unit glass area (Table H1)	Total solar load (W) allowing for 10% framing
South	5 x 2.8 x 0.65 = 9.10	156	= 9.10 x 156 x (1-0.1) = 1278
West	3 x 2.8 x 0.65 = 5.46	205	= 5.46 x 205 x (1-0.1) = 1007
Total solar load	= 1278 +1007 = 2285		

In this case S_c = (0.7 x 25 x 5 x 3)/2285 = 0.11. Such a shading coefficient is quite demanding to achieve in practice. One alternative might be to reduce the glazing area by using opaque insulating panels. If the building is to be air conditioned, another option would be to demonstrate that the Carbon Performance Rating of the building's ACMV systems is no greater than the values shown in Table 11.

If the corner office was not partitioned from a general open floor area, it's solar load could be considered as part of the load of one of the facades it shares.

[2] **BS EN ISO 13789:1999** Thermal performance of buildings - Transmission heat loss coefficient - Calculation method

[3] **BS EN 12664:2001** Thermal performance of building materials and products – Determination of thermal resistance by means of guarded hot plate and heat flow meter methods – Dry and moist products of low and medium thermal resistance

[4] **BS EN 12667:2000** Thermal performance of building materials and products – Determination of thermal resistance by means of guarded hot plate and heat flow meter methods – Products of high and medium thermal resistance

[5] **BS EN 12939:2001** Thermal performance of building materials and products – Determination of thermal resistance by means of guarded hot plate and heat flow meter methods – Thick products of high and medium thermal resistance

[6] **BS EN ISO 8990:1996** Thermal insulation – Determination of steady-state thermal transmission properties – Calibrated and guarded hot box

[7] **BS EN ISO 12567-1:2000** Thermal performance of windows and doors – Determination of thermal transmittance by hot box method – Part 1: Complete windows and doors

[9] **BS EN ISO 6946:1997** Building components and building elements – Thermal resistance and thermal transmittance – Calculation method

[10] **BS EN ISO 13370:1998** Thermal performance of buildings – Heat transfer via the ground – Calculation methods

[11] **BS EN ISO 10077-1:2000** Thermal performance of windows, doors and shutters – Calculation of thermal transmittance – Part 1: Simplified methods

[12] **prEN ISO 10077-2** Thermal performance of windows, doors and shutters – Calculation of thermal transmittance – Part 2: Numerical method for frames

[19] **BS EN ISO 10211-1:1996** Thermal bridges in building construction – Calculation of heat flows and surface temperatures – Part 1: General methods

[20] **BS EN ISO 10211-2:2001** Thermal bridges in building construction – Calculation of heat flows and surface temperatures – Part 2: Linear thermal bridges

[21] **BS EN 12524:2000** Building materials and products – Hygrothermal properties – Tabulated design values

[26] **BS EN ISO 13789:1999** Thermal performance of buildings - Transmission heat loss coefficient - Calculation method

[41] **BS 5422:2001** Method for specifying thermal insulating materials for pipes, tanks, vessels, ductwork and equipment operating within the temperature range –40°C to +700°C

[55] **BS 7913:1998** The principles of the conservation of historic buildings

L2 OTHER PUBLICATIONS REFERRED TO

The Stationery Office

[1,27] Limiting thermal bridging and air leakage: Robust construction details for dwellings and similar buildings, TSO 2001

[47] Guidelines for environmental design in schools, Building Bulletin 87, TSO, 1997.

[48] NHS Estates: Achieving energy efficiency in new hospitals, TSO, 1994.

British Cement Association (BCA) and National House Building Council (NHBC)

[12,21] Approved Document – Basements for dwellings, 1997, ISBN 0-7210-1508-5

Building Research Energy Conservation Support Unit (BRECSU)

[34] GIR 31, 1995 Avoiding or minimising the use of air-conditioning

[38] GPG 303, 2000 The designer's guide to energy-efficient buildings for industry.

[39] GPG 132, 2001 Heating controls in small commercial and multi-residential buildings

[45] GIR 41, 1996 Variable flow control

[53] GIL 65, 2001 Sub-metering New Build Non-domestic Buildings: A guide to help building designers meet Part L of the Building Regulations.

Building Research Establishment (BRE) (published by CRC Ltd)

[8] Conventions for the calculation of U-values: expected publication date early 2002

[14] IP 5/98 Metal cladding: assessing thermal performance

[18] U-value calculation procedure for light steel frame walls: expected publication date early 2002

[28] IP 17/01 Assessing the effects of thermal bridging at junctions and around openings

[30] BR 262, 2002 Edition Thermal insulation: avoiding risks,

[33] BR 364, 1999 Solar shading of buildings

[43] IP 2/99 Photoelectric control of lighting: design, set-up and installation issues

[46] Digest 457, 2001, The Carbon Performance Rating for offices,

[50] BR 176, 1991 A practical guide to infra-red thermography for building surveys,.

Centre for Window and Cladding Technology (CWCT)

[16] Guide to good practice for assessing glazing frame U-values (1998, new edition in preparation)

[17] Guide to good practice for assessing heat transfer and condensation risk for a curtain wall (1998, new edition in preparation)

Chartered Institution of Building Services Engineers (CIBSE)

[22] Guide A: Environmental design, Section A3: Thermal properties of building structures, 1999

[24] Energy efficiency in buildings, 1998

[31] LG10: Daylight and window design, 1999

[35] Guide A: Environmental Design, Section A5: Thermal response and plant sizing, 1999.

[40] Guide H: Building Control Systems, 2000

[44] TM22: Energy Assessment and Reporting Methodology: Office Assessment Method, 1999

[49] AM11; Building Energy and Environmental Modelling, 1998

[51] TM23: Testing buildings for air leakage, 2000

[54] Energy Efficiency in Buildings, Chapter 11: General electric power, 1998

Commissioning Specialists Association

[52] Technical Memorandum 1: Standard specification for the commissioning of mechanical engineering services installations for buildings, 1999

Council for Aluminium in Building (CAB)

[15] Guide for assessment of the thermal performance of aluminium curtain wall framing, 1996

Department of the Environment (DoE) and Department for National Heritage (DNH)

[23] Planning and the historic environment, Planning Policy Guidance PPG 15, DoE/DNH, September 1994. (In Wales refer to Planning Guidance Wales Planning Policy First Revision 1999 and Welsh Office Circular 61/96 Planning and Historic Environment: Historic Buildings and Conservation Areas.)

Department of the Environment, Transport and the Regions (DETR)

[25] Energy Consumption Guide 19: Energy use in offices-, 1998

[37] CHPQA Standard: Quality Assurance for Combined Heat and Power, Issue 1, November 2000.

Metal Cladding and Roofing Manufacturers Association (MCRMA)

[29] Technical Note 14: Guidance for the design of metal cladding and roofing to comply with Approved Document L 2002 Edition:

Society for the Protection of Ancient Buildings (SPAB)

[56] Information sheet 4: The need for old buildings to breathe, 1986.